DIGEST

A Primer for the International GIS Standard

DIGEST

A Primer for the International GIS Standard

Kelly Chan, Ph.D.

Environmental Systems Research Institute (ESRI)
Redlands, California

Lewis Publishers

Boca Raton   Boston   London   New York   Washington, D.C.

Library of Congress Cataloging-in-Publication Data

Chan, Kelly.
    Digest : a primer for the international GIS standard / Kelly Chan.
      p. cm.
    Includes index.
    ISBN 1-56670-241-0 (alk. paper)
    1. Geographic information systems--Standards.
  G70.212.C47   1998
  910′.285--DC21   98-5981
                                                                 CIP

© 1999 by CRC Press LLC
Lewis Publishers is an imprint of CRC Press LLC

No claim to original U.S. Government works
International Standard Book Number 1-56670-241-0
Library of Congress Card Number 98-5981
Printed in the United States of America 1 2 3 4 5 6 7 8 9 0
Printed on acid-free paper

Preface

In the late 1980s, the National Imagery and Mapping Agency (NIMA — then known as the Defense Mapping Agency) had taken a major step into the world of Geographic Information Systems with a multinational program known as the Digital Chart of the World (DCW). Military mapping agencies from Australia, Canada, the United Kingdom, and the U.S. participated in the DCW project.

A major goal and one of the most significant outcomes of the DCW project was the development of a generic, machine-independent format for geospatial data to be used to support military analysis. This format, the Vector Product Format, or VPF, was implemented for the DCW database and other product series.

The goals for the VPF were to support GIS applications, to be compatible with the Digital Geographic Information Exchange Standard (DIGEST), and to directly support data quality information. The VPF was based on a georelational model similar to ones used in several commercial GIS formats at that time and was intended to support direct data use in GIS applications — not to be only a data exchange format standard. Another aspect of the VPF design was to include data quality information to permit users to evaluate the suitability of the data for a particular application.

The driver behind the development of VPF was NIMA. NIMA was formed in 1996 with consolidation of U.S. national and DoD imagery and mapping organizations. The objective of the consolidation was to exploit current and future collection systems and digital processing technologies and provide timely support to their users. NIMA has the responsibility to provide imagery, imagery intelligence, and geospatial information to U.S. DoD and national users. The former Defense Mapping Agency (DMA) was part of the reorganization. DMA was created in 1972 with the consolidation of the U.S. armed services mapping organizations. The Digital Geographic Information Working Group, DGIWG, was organized to establish digital geographic data standards that can be used by digital spatial data producer nations and users to exchange data and support interoperability within the NATO community. The member nations of DGIWG include Belgium, Canada, Denmark, France, Germany, Italy, the Netherlands, Norway, Spain, the United Kingdom, and the U.S. DGIWG produced the standard, Digital Geographic Exchange Standard (DIGEST), which defines a set of rules and encoding conventions for the exchange of spatially referenced raster, vector, and matrix data. VPF was designed to be compatible with DIGEST. In 1991, the U.S. Department of Defense Vector Product Format (VPF) was incorporated, with minor changes, into the DIGEST version 1.1 as the Vector Relational Format (VRF).

The development of the VPF standard and the DCW database was conducted in three phases under the DCW project: design, prototyping, and data production. The VPF standard emerged from the design and prototyping phase and was implemented in the DCW production phase. The design and prototyping phase lasted from October 1989 through January 1991. Four prototypes were built to evaluate the VPF and DCW product designs and establish data automation procedures for implementing the VPF. The VPF standard evolved throughout the prototyping process. Each prototype was an interim product with participants reviewing and commenting on the design. The VPF standard was completed in the spring of 1991. Subsequent reviews by the international participants resulted in a series of Request-For Changes to the standard. In April 1992, NIMA released the Vector Product Format Military Standard (MIL-STD-600006) version 1.0. The VPF standard has continued to evolve as experience and use become more prevalent. The VPF Military Standard version 1.1 was released in 1993 and was the base standard used for many production efforts by NIMA. The standard was again reissued in 1996 to accommodate better definition of cross tile topology.

An objective of the VPF standard development was to ensure that the standard could be used for other products besides the DCW. The DCW was the initial implementation of the VPF standard. Six other prototype developments were conducted under the DCW program.

DIGITAL CHART OF THE WORLD (DCW)

The first database produced using the Vector Product Format was the DCW. The DCW contains data equivalent to a 1:1,000,000 scale map and was based on the content of the Operational Navigation Charts. The DCW database includes information covering the entire earth. It contains 1700 megabytes of geographic data in 17 thematic layers and is on 4 CD-ROM disks.

VECTOR SMART MAP LEVEL 0 (VMAP LEVEL 0)

The VMap Level 0 database is a revised/updated version of the DCW, and the features are divided into 10 thematic layers similar to other products in the VMap series (VMap Levels 1 and 2 and Urban Vector Smart Map, UVMap). As with the DCW, the VMap Level 0 provides low resolution (equivalent to 1:1,000,000 scale) vector-based data. VMap Level 0 thematic layers (boundaries, elevation, hydrography, industry, physiography, population, transportation, utilities, vegetation, and data quality) are consistent throughout the VMap program. In addition, low resolution bathymetry is added to the Level 0 data. In contrast to the DCW, the VMap Level 0 feature coding scheme is based on the Feature/Attributre Coding Catalog (FACC), developed by the Digital Geographic Information Working Group (DGIWG) to support the Digital Geographic Information Exchange Standard (DIGEST).

DIGITAL NAUTICAL CHART (DNC)

The original requirement for the Digital Nautical Chart (DNC) was to support computerized marine navigation and Geographic Information System (GIS) applications. DNC is a vector database of selected maritime features. The DNC databases are organized into libraries based on the data sources used (Harbor, Approach, Coastal, and General charts). The data are organized into 12 thematic layers: cultural landmarks, earth cover, environment, hydrography, inland waterways, land cover, limits, aids to navigation, obstructions, port facilities, relief, and data quality.

INTERIM TERRAIN DATA (ITD)

Interim Terrain Data was an interim product to support digital terrain analysis prior to development of the Tactical Terrain Data (TTD). The ITD data was based on the content and scale of the Tactical Terrain Analysis Data Bases (TTADBs) or Planning Terrain Analysis Data Bases (PTADBs) with an enhanced transportation file. The data sets are composed of feature information and attributes. Interim Terrain Data is frequently used with Digital Terrain Elevation Data (DTED) and raster maps, such as the ARC Digitized Raster Graphic (ADRG) base maps.

DIGITAL TOPOGRAPHIC DATA (DTOP)

DTOP contains data in thematic layers from both terrain analysis and topographic line map products. The DTOP is the VPF implementation of the terrain analysis portion of Tactical Terrain Data (TTD). TTD will provide terrain information for planning and operational uses such as terrain visualization, mobility/countermobility planning, site/route selection, reconnaissance planning, communications planning, navigation, and munitions guidance close air support missions, amphibious operations, and land combat operations. Currently, the DTOP is in the prototyping phase. It is intended that the TTD will consist of a combination of Digital Topographic Data (DTOP), Digital Nautical Chart (DNC), and Digital Terrain Elevation Data Level 2.

WORLD VECTOR SHORELINE (WVS)

The World Vector Shoreline (WVS) is a digital database of international boundaries, shorelines, and country names. The World Vector Shoreline is a standard NIMA product used to support global analysis and display applications.

VECTOR SMART MAP LEVEL 1 (VMAP LEVEL 1)

The VMap Level 1 data is a medium resolution vector-based geospatial data set. The VMap Level 1 data are divided into the standard VMap thematic layers (boundaries, elevation, hydrography, industry, physiography, population, transportation, utilities, vegetation, and data quality) and reference layer. The data were originally derived from 1:250,000 source maps, but are also now being produced from photogrammetric sources.

VECTOR SMART MAP LEVEL 2 (VMAP LEVEL 2)

The VMap Level 2 is a high-resolution georelational vector database with an equivalent resolution of 1:50,000 to 1:100,000. The data are organized into the standard VMap thematic layers and are organized into libraries based on geographic areas.

URBAN VECTOR SMART MAP (UVMAP)

The UVMap is attributed vector-based data of cities and is based on the NIMA city graphic and military city map products. UVMap is centered around specific cities.

The primary aim of this book is to help orient the reader to the VPF geospatial data standard. The hope is that this will expose the reader to valuable geospatial data.

FOUNDATION FEATURE DATA (FFD)

NIMA has developed a foundation feature data prototype specification and VPF product. The FFD contains key boundaries, elevation, surface drainage, population, transportation, and vegetation information. This data base will form a base or "foundation" set of data which users can enhance or value-add feature content or resolution for their specific requirements. The concept is that the FFD contains basic essential data for broad areas. These data are the base or control for users to add the mission-specific data.

About the Author

Kelly Chan, Ph.D. is a senior member of the technical staff with the Professional Services Division of ESRI based in Redlands, California. He is responsible for the strategic planning, integration, and application of information management technologies and systems for ESRI clients and has applied GIS database and spatial analytical techniques to a variety of project activities. He was the VPF manager for the Digital Chart of the World project.

Dr. Chan received his B.A. (honors) and master's degrees in geography from the University of Western Ontario, Canada, and his doctorate in urban planning from Harvard University. He was a research associate with the Laboratory for Computer Graphics and Spatial Analysis in the Harvard Graduate School of Design.

Acknowledgments

Many people at ESRI worked on the DCW project. The VPF Standard was essentially completed by the time I joined the project.

David Flinn was the original architect responsible for the Standard work.

Many thanks to Lee Peterson for his support and encouragement.

Thanks to Jack Dangermond, President of ESRI, for support and for allowing the use of materials in this book, and to Momay, who helped finish it.

Contents

CHAPTER 1

VPF and DIGEST/VRF

This primer is prepared for two military standards: (1) U.S. Department of Defense, Mil-Std-2407, and (2) the North Atlantic Treaty Organization's STANAG 7074, DIGEST.

The complete title for Mil-Std-2407 is "Interface Standard for Vector Product Format." The revision referenced in this book was released in June, 1996. STANAG 7074 refers to "Standardization Agreement, Digital Geographic Information Exchange Standard." The version of DIGEST used here is version 1.2a, published in May, 1995.

The aim of this primer is to introduce the readers to a vector format utilized by these agencies for geographic data. In that aim, these two documents are essentially identical, although DIGEST covers a slightly broader collection of data formats which includes non-vector-based formats.

In addition, ISO Technical Committee, ISO/TC 211, Geographic Information/Geomatics identifies DIGEST as one of the *de facto* standards to be incorporated into the emerging ISO standards under this efforts. This primer will be applicable beyond the military sphere in the near future.

In his report in the ISO bulletin, ISO/TC 211, chairman Olaf Ostensen wrote,

> The arms trade is international, and so, too, are more and more paramilitary operations like peace-keeping operations under UN command. It is, thus, no wonder that the military has addressed the GIS standardization issues for the last two decades. The most mature results are with Digital Geographic Information Working Group (DGIWG), a NATO-based group, which has defined the DIGEST family of geomatics standards.
>
> While there is normally a longish gap between the definition of standards and their real and practical implementation, the DGIWG countries have also worked hard at producing data conforming to the standard. Thus, already a couple of years ago, we had a global dataset of geographic information, Digital Chart of the World, delivered on four CD-ROMs conforming to DIGEST. Although the resolution of the data is somewhat coarse, corresponding to maps in scale 1:1 000 000, this is the first example of a general geographic dataset covering the whole world. The dataset is furthermore available at close on distribution cost, and several system vendors have tailored it to their own systems. At present, the same group is at work on a global dataset representing mapping

to the scale of 1:250,000 to be available by the end of the century. This dataset will fill 234 CD-ROMs, that is to say about 150 GB of information!

ISO/TC 211's liaisons include DGIWG, that is also aiming to harmonize its present *de facto* standards with the future ISO standards as profiles. If we manage to bring about this harmonization, we will have an International Standard that is already accepted by the user community and with a vast amount of information in compliance with the standard. This would really herald a flying start for ISO/TC 211 Standards.*

Ostensen's report was written in December, 1995. A vast volume of data has since been published in compliance to the Mil-Std-2407 and DIGEST.

As shown in the title of Mil-Std-2407, Vector Product Format (VPF) is the preferred term for the U.S. Department of Defense. NATO, however, has chosen the name of Vector Relational Format (DIGEST/VRF). These two terms are used interchangeably here. For the most part, the shorter form VPF is used in this book.

INTERFACE STANDARD FOR VECTOR PRODUCT FORMAT

Mil-Std-2407 was revised and released on June 28, 1996. The document is almost 300 pages long, whereas the standard specifications themselves take up only about 100 pages. The remaining pages of the document include 8 appendices. Each of these appendices covers one specific topic area. For example, the first presents a justification statement and a brief design philosophy behind the development of VPF. By and large, these appendices are mostly examples rather than in-depth discussions.

Of particular interest is perhaps Appendix H, which contains a formatted text printout of a VPF database. Although the printout is presented without an accompanying explanation text, the example is, nonetheless, an extremely valuable introduction to a user trying to get a first-hand understanding of the VPF standard.

Appendix G includes a number of coding schemes which are not presented elsewhere in the document. Since the VPF standard focuses solely on a format, variable coding schemes are generally not treated as parts of the standard specifications. Variable coding schemes are included in customized product specifications, which describe particular geographic data products published in compliance with the VPF standard, such as the aforementioned Digital Chart of the World. The coding schemes shown in this appendix are used to encode meta data attributes. Since meta data entries are common to all products and, in fact, should even be shared by datasets published in formats other than VPF, it is only logical that a standardized coding scheme be included in the VPF document for meta data.

MIL-STD-2407 — ORGANIZATION

The organization of Mil-Std-2407 is prepared in accordance with other military standards which govern the layout and content of documents of this type.

* Ostensen, Olaf "Mapping the future of geomatics," ISO Bulletin, December, 1995.

NOT MEASUREMENT
SENSITIVE

MIL-STD-2407
28 JUNE 1996
SUPERSEDING
MIL-STD-600006
27 APRIL 1996

DEPARTMENT OF DEFENSE

INTERFACE STANDARD FOR
VECTOR PRODUCT FORMAT

AMSC N/A   AREA MCGT

Figure 1.1 VPF cover.

The first section of the standard is the scope statement. Section 2, "Applicable Documents," enumerates the relevant documents and standards. Of particular interest are the ISO standards regarding character encoding.* In addition, there are the ANSI/IEEE encodings for floating point numbers. Included in this list of applicable

* These encodings include ISO 646, ISO 2022, ISO 2375, ISO 6937, ISO 8859.1 and ISO 10646-1.

documents is the NATO STANAG 7074, DIGEST version 1.2. As noted earlier, DIGEST has since been updated to version 1.2a.

Even though these documents are prominently listed at the beginning of the standard, the "order of precedence" statement at the end of this section explicitly notes that

> In the event of a conflict between the text of this document and the references cited herein, the text of this document takes precedence...

A list of definitions is presented in Section 3. This is a glossary of sorts. This glossary begins on page 4 and ends at page 15. Although most definitions are only a one- or two-sentence short description that might not be sufficient to most readers, this section provides a good navigational aid.

The general requirements outline the VPF standard and its applications. Figure 1, on page 17, shows the relationships between VPF the standard, which defines a format, and specific VPF products that are published using the format.

Demanding special attention in this section is the relationship between the VPF standard and product specifications that describe products constructed in compliance with the VPF standard.

The purpose of this section is to make clear the distinctions between the contents of a geographic database; the form, which is VPF, in which the contents are encapsulated; the resultant data products; and their applications and end users.

The standard specifications actually start at Section 5, "Detailed Requirements." This section is itself divided into three smaller parts. The first subsection describes the data model. It is followed by an implementation subsection which shows how the data model is implemented using relational database concepts of data columns, records, and tables. Finally, the last subsection describes how the columns, records, and tables are encoded to achieve the mandates of the standard: a direct-use and exchange format for geographic data.

The data model in VPF is presented in Section 5.2. The discussions begin at page 20. Figure 4, on page 24, graphically captures the structural levels of: (1) feature class, (2) coverage, (3), library, and (4) database in a VPF database.

VPF data model components are introduced in this section, beginning with primitives in Section 5.2.2.1, at page 27. Primitives are the basic geometric building blocks, the alphabets, which give the graphical representation of geographic data in a VPF database.

The geographic meanings of these data are modelled by feature classes, which are introduced in Section 5.2.2.2. Geographic data are organized into containers called coverages. These are the subjects of Section 5.2.2.3.

Coverages are not only containers where a database developer can efficiently organize data in a database to maximize data usability. They also support design considerations such as spatial integrity constraints and physical partitioning. The brand of integrity constraints enforced in a VPF database is called topology, which is discussed in Section 5.2.2.3.1 in the standard. VPF databases can be partitioned for enhanced processing efficiency by means of tiling. Section 5.2.2.3.3 of the standard discusses tiled coverages. Cross-tile keys are designed to recover a tiled coverage into its original whole.

The notion of data quality and the mechanisms available to encode data quality information in a VPF database are presented in Section 5.2.3. This completes the data model subsection.

The implementation subsection, Section 5.3, shows a relational database implementation of the data model just presented. This subsection parallels the discussions in Section 5.2. Some particularly noteworthy discussions include the tables of reserved names in Section 5.3.1.2.

The implementation for primitives is presented in Section 5.3.2. It is followed by the discussions on feature classes, Section 5.3.3. How a feature gets its geometric definitions is shown in the feature join table implementation, Section 5.3.3.2. To handle a partitioned dataset, Section 5.3.3.3 describes the implementation of feature-to-primitive relations in tiled coverages.

Two implementation changes that might not be familiar to users of previous revisions of the standard are found in Sections 5.3.5.5 and 5.3.5.6. Section 5.3.5.5 shows the new implementation of the registration point table, while Section 5.3.5.6 shows that of the diagnostic point table. Both of these tables are related to the definition of a coordinate reference system for a library in a VPF database.

Other than the organizational containers of libraries and coverages, everything in a VPF database is stored in VPF tables. Section 5.4, "VPF Encapsulation," describes how these tables are constructed. Table definition is presented in Section 5.4.1. Besides these tables, this subsection also introduces structures of various indices that have been designed to accompany the tables in VPF.

Another area of capsulation deals with the naming convention established in VPF. Naming convention is discussed in Section 5.4.5.

Most VPF data types are straightforward data types easily recognized by those who are familiar with commercially available relational databases. One such type is the notion of coordinate strings which are composed of repeating floating point numbers. Perhaps the only unusual data type is the triplet id. The triplet id field type is described in Section 5.4.6.

It is hoped that this brief tour of Mil-Std-4027 will help the reader of the primer locate relevant subjects in the standard document. Mil-Std-4207 is attached in DIGEST, almost verbatim, as Section 2, Annex C. There are, however, other areas in DIGEST that a reader of this primer should pay close attention to.

BASIC CONCEPTS

There are a number of key terms frequently used in the Standard. Of particular interest are those concerned with the Standard and compliance with the Standard. The VPF and DIGEST/VRF Standard are, for practical purposes, identical documents. Every effort by the various agencies involved has been to consolidate the two documents into one. The Standard specifies a set of mandatory tables and columns in these tables. In addition to the mandatory ones, there are also optional tables and columns. There is no restriction on the introduction of other tables and columns into a database published to the Standard.

The

**DIGITAL GEOGRAPHIC INFORMATION
EXCHANGE STANDARD**

(DIGEST)

Support Document:

Detailed Logical Flow Diagrams

Produced and issued under the direction of the Directorate of Geographic Operations,
Department of National Defence, Canada on behalf of the
Digital Geographic Information working Group.

Edition 1.2   January 1994

Figure 1.2   DIGEST/VRF cover.

Any spatial database published according to the Standard must have a product specification that clearly states which of the optional tables and columns are included and completely define all other tables and columns not specifically described in the Standard.

A product specification is considered to be in compliance with the Standard if it includes all mandatory tables and columns, and the product-specific tables and columns do not alter the interpretation of the mandatory and optional tables and columns as stated in the VPF and DIGEST/VRF Standard.

A VPF compliant application is one that properly interprets all mandatory and optional tables and columns as stated in the Standard.

The following is excerpted from Section 5.2.1.3 of the Interface Standard for Vector Product Format:

> This document describes the column definitions for all the VPF standard-specified columns, and the table organization for those columns. Data columns and tables described in this document are labeled either mandatory or optional. A VPF product must include all mandatory tables and columns. It is not possible to remove any mandatory column from any table. A VPF-compliant application must be able to process a VPF product and interpret all mandatory and optional columns as described in this document.
>
> Additional product-specific columns are allowed in VPF. If present, these columns must be defined in their product specifications. Product-specific columns must not alter the use of the columns specified in this document.

VPF is defined using a relational data model. This model is called a georelational data model because of the spatial or geometric attributes that always participate in every relation. These geometric attributes are represented in a set of geometric primitive tables. A geographic feature is constructed by joining one of these geometric primitive tables with a feature base table.

These feature relationships are governed by conventional relational integrity constraints. This is no different from any other traditional database operation.

> Geometric primitives are, in addition, governed by a special set of integrity constraints known as topological integrity constraints. This type of constraint is common in most geospatial databases. The VPF implementation of topological integrity constraints is based on a data structure called winged-edge topology.

CHAPTER 2

A Database Tour

The best introduction to a VPF database is by examining some sample databases, Figure 2.1 shows a directory view of a sample VPF CD-ROM. This CD-ROM was first published by the U.S. Defense Mapping Agency, as it was then known, in 1995. The first figure shows the three sample databases: (1) dtopn, (2) vmaplv0, and (3) vmaplv2. This series of screens were taken from Windows NT™. The sample CD-ROM has been written to the ISO-9960 standard and is accessible under other operating systems, such as UNIX™, Windows™, and the Macintosh OS™.

Figure 2.2 continues down the folders and now shows the contents of the dtopn database. Inside this database there are two libraries: (1) rference and (2) t101us01. The exact details of these libraries are defined in the appropriate product specifications; in this case dtopn is described by the MIL-PR-89037, Digital Topographic Data. The reference library is a thumbnail description of the database. This library provides content information about the actual database. Since this is only a sample data disk, the *library* t101us01 is only a very small subset of the entire dtopn database. Many VPF databases, especially those with worldwide coverages, are subdivided into geographic libraries. T101us01's area of interest covers a relatively small region in the continental U.S.

The actual data content of this sample dataset is in *library* t101us01. Figure 2.3 shows the items found at the library level. There are 19 items here. There are six meta-data files. These meta-data files describe the content of the *library*. VPF databases are self-documenting. Viewers, such as the two shown below, can comfortably access and retrieve data from a database based on these meta-data elements. Meta-data are not restricted to simple tables or files. Among the 13 subdirectories, there are two meta-data subdirectories. The libref is a reference *coverage* similar to the rference library, however, libref gives a thumbnail description of the *library* rather than that of the *database*. The tileref describes the physical storage structure of the database. Eleven of the other subdirectories are *coverage* subdirectories. These contain the actual data.

Fortunately, using a VPF database seldom means navigating this directory information directly. Figure 2.4 shows how to access these data in ArcView™ using the

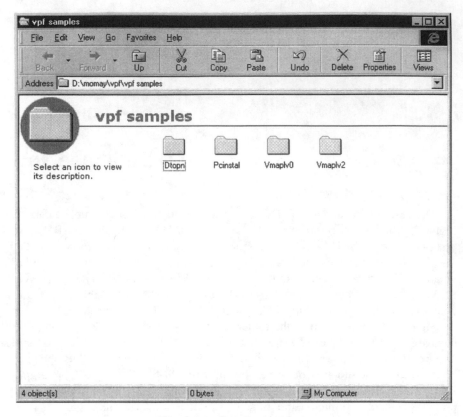

Figure 2.1   Sample databases.

VPF Viewer extension. The ArcView GIS Version 3.0™ is a product by Environmental Systems Research Institute.

Figure 2.4 shows how to navigate down a disk to the t101us01 library. Instead of seeing the thirteen *coverage* directories, the list of *coverages* is shown in more descriptive names. Returning to an earlier figure, Figure 2.3, this information is actually stored in the *coverage attribute table*, cat, file.

Navigating further down the *dataset*, a *coverage* is made up of various feature classes. Under the *surface drainage*, sdr, *coverage* there are eight *feature classes*. Figure 2.5 shows some of these in the scrolling list on the left-hand side of the add theme dialog in ArcView.

Figure 2.6 shows a map display of the selected *features*. It shows the various *area* and *line features* in the *surface drainage coverage*. In addition to the *feature classes* in this one *coverage*, the map view also includes a neat line box from the *tile reference coverage*. In a VPF database, all *coverages* in a *library* are assumed to share the same coordinate reference system. Hence, it is obvious that *features* from different *coverages* in a *library* can be displayed simultaneously. A map can be constructed using different combinations of *feature classes* from separate *coverages* in a *library*, as Figure 2.6 shows.

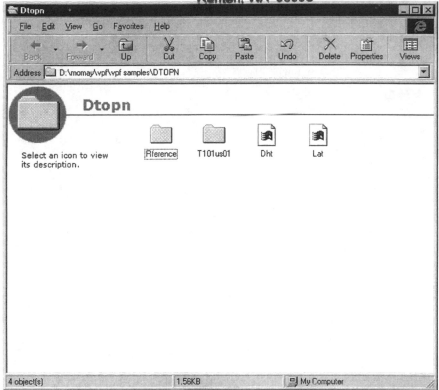

Figure 2.2 VPF database dtopn.

Figure 2.7 shows an attribute table of the *lake feature class*, again in ArcView. The *lake feature class* is one of eight *feature classes* in the *surface drainage coverage*. ArcView contains a rather comprehensive collection of viewing and browsing functions that are not restricted to VPF data. Besides browsing functions, there are also other analytical tools that allow the use of VPF data in various operations such as land management analysis and logistic planning. The objective of the Vector Product Format is to support this direct use of spatial data.

Figure 2.8 shows the library content of sdr, the *surface drainage coverage*. What has been hidden from us by ArcView is quite a large number of files and directories that actually store the data, both graphical lines and polygon shapes, and corresponding attributes that allow us to construct the simple map display in Figures 2.6 and 2.7.

To summarize, a VPF *database* contains multiple *libraries*. In this example, the dtopn *database* is composed of two *libraries*: (1) rference and (2) t101us01. *Libraries* can be organized by data content. For example, rference provides a brief, index type description and t101us01 contains the actual detailed data. *Libraries* can also be organized by geographic area: t101us01 covers part of the U.S.; perhaps another t101uk01 covers the United Kingdom. A *library* is made up of *coverages. Coverages* are thematic organization. There is a *coverage* for *surface drainage*; and a coverage for *transportation* and *vegetation*.

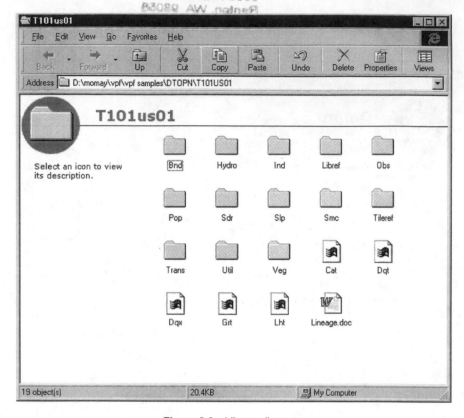

Figure 2.3   Library directory.

The Windows NT active desktop shows these files and directories with a capitalized first character, but the actual files and directories use all lowercase names.

There are other software packages and commercial products that are capable of handling VPF datasets besides the ArcView. Figure 2.9 shows a public domain software, the VPFViewer 2.0, which is usually distributed along with every product CD-ROM. VPFViewer is published by the U.S. National Imagery and Mapping Agency.

ANNOTATED EXAMPLE OF A VPF DATABASE

A rather unusual VPF database is presented in the next few pages. This is not a real database such as the one shown earlier. In fact, no data have ever been created for this database and no product specification has been written for it. This database is unusual for two reasons. First, the database probably includes everything possible in the VPF Interface Standard. A few of these items most likely will never be encountered in true production, as released databases. Second, our example database contains an exceedingly large number of tables, yet fewer than 10 *feature classes* are defined. (Remember that *feature classes* are our vocabulary for modelling geography using VPF. Let us hope this aspect of the example remains unusual!) Even

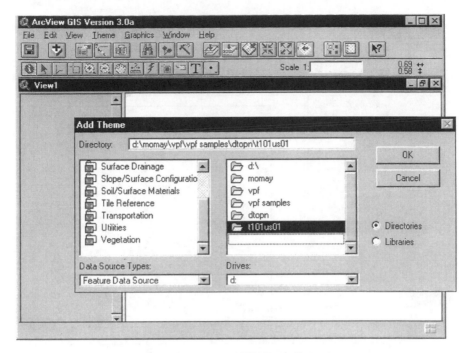

Figure 2.4   ArcView VPF Viewer Extension.

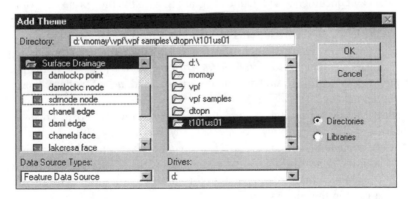

Figure 2.5   Arcview, Add Theme dialog.

though it is peculiar, this database gives a reasonably comprehensive orientation of DIGEST/Vector Relational Format and the Interface Standard for Vector Product Format in general.

Again, let assume that an end-user inserts and mounts a disk containing the sample VPF database. Without a browser such as ArcView, the VPF database folder resembles a directory listing in Figure 2.10. Figure 2.10 shows folders and documents. Folders represent directories and subdirectories, and documents are, by and large, files that stores VPF tables. By definition, one file stores one table in VPF.

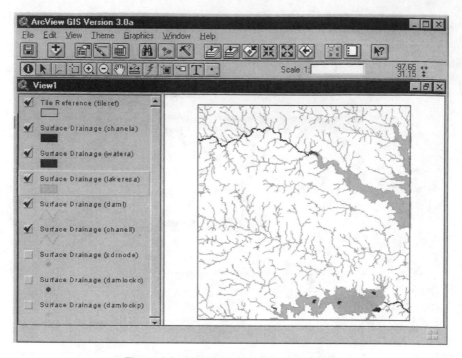

Figure 2.6   VPF data displayed in ArcView.

Folders and documents with names beginning with the my_ prefix are, of course, mine. These are items made up for this example. They are used to store specific domain data in a VPF database. This convention is purely to help in our guided tour here to distinguish what are domain-specific and what are required by the Standard. A domain data element might be the *surface drainage coverage*, sdr, in the previous example. Otherwise, folders and documents are specified in the Standard; their names are reserved names.

Figure 2.10 shows a database directory of the sample *database*. This *database* has four *libraries*; they are: my_lib, my_lib1, my_lib2 and my_lib3.

The dht document contains the *database header table*. This table provides information about the *database*. The lat document stores the *library attribute table*. The four *libraries* in the *database* are listed in this lat table with more descriptive names than those used in the file system.

These two tables are required, that is, they appear in all databases. The user is also likely to encounter the *data quality table*, dqt, and an accompanying documentation table, lineage.doc. The *data quality table* maintains data quality reports that are increasingly becoming mandatory according to established data library and transfer standards. Lineage generally describes the development history of the database; this includes data source and production methods.

Other descriptive tables, called narrative tables, are also likely. In this example my_story.doc is one of these narrative tables. A narrative table can be attached to any other table or any other data field in a table. They are simply a means to associate descriptive narratives with database elements.

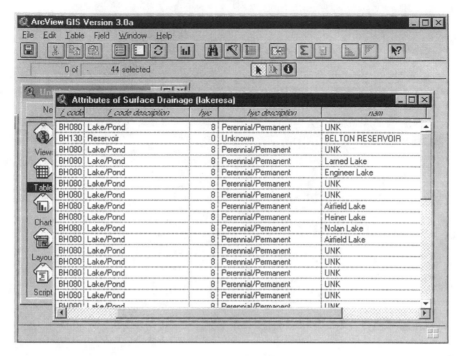

Figure 2.7   VPF attribute data shown in ArcView table view.

One will no doubt also note the large number of pesky tables with names that end in an x. These are index tables. Therefore, dhx is an index for dht, my_story.dox is an index for my_story.doc, and so on.

Figure 2.11 shows the contents in the my_lib *library* of the sample *database*. It is obviously parallel in structure to the database level in Figure 2.10. Instead of the *database header table*, there is the *library header table*, lht. And rather than *libraries*, this a *coverage attribute table*, cat, which in turn enumerates the four *coverages* in this *library*. Without being too creative, the four *coverages* are named: my_cov, my_cov1, my_cov2 and my_cov3.

Data quality tables are here also. It is possible to include data quality at all levels in a VPF database. Data quality and lineage information might be different at different levels and for different regions.

Recall the earlier example where *features* from different *coverages* are displayed on a single map. Coordinate reference systems are defined for *libraries*. This is defined using three tables, the *geographic reference table*, grt, the *registration points table*, my_lib.rpt and the *diagnostic points table*, my_lib.dpt. The *geographic reference table* defines a projection model, usually by specifying a set of standard projection parameters. Depending on the method the data are captured for a *library*, the *registration* and *diagnostic points* are for validating the source materials and the projection model.

Besides the four *coverage* subdirectories, there are four other subdirectories. The *tile reference coverage*, tilref, defines a storage scheme for graphics in this library.

Figure 2.8   Directory content of a coverage.

The *library reference coverage*, libref, is essentially a thumbnail representation of the detail contents in the *library*. There is also the *data quality coverage*, dq, which is a companion of the *data quality tables*. This *coverage* is sometimes referred to as the data quality overlay. It graphically outlines data quality variations over different areas within the geographical extent of the *library*. Finally, the gazette *coverage* is a geographic index. Remember looking up a place in a dusty atlas at the public library? That is a gazette. A gazette stores prominent place names, and in its printed version the gazette points one to the page and a box in which the place is found. It behaves pretty much the same in the digital version in VPF. Given a name, one can quickly locate where the place is found in a VPF library.

Do note that a number of documents in the *library* subdirectory share the same name as their counterparts in the earlier *database* subdirectory. It is perfectly acceptable in VPF. For example, *data quality tables* are found in both levels, and perhaps in a few more later on.

Both reserved names and domain specific database names can be reused in different contexts. Hence, my_story.doc is retold many times in the databases, possibly with a different story each time.

Figure 2.12 shows the contents of the *library reference coverage*, libref. As introduced earlier, this reference *coverage* is a simplified "snapshot" of the detailed

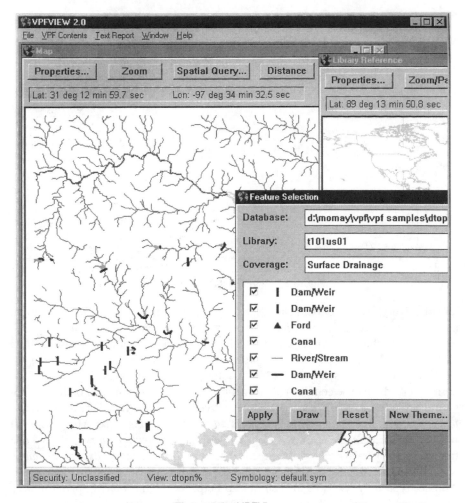

Figure 2.9 VPFViewer.

content of the *library*. *Library reference coverages* are generally very simple *coverages*, in that they allow browsers to quickly present a preview of a dataset. End-users can thus decide whether to investigate the dataset further.

This simplicity also provides us an excellent opportunity to tour the content of a *coverage*. Figure 2.12 shows 11 documents in the *library reference coverage* subdirectory. Perhaps the most important document shown in Figure 2.12 is the *feature class schema table*, fcs. This is the usual case in all *coverages*.

The *feature class schema table* defines how things are put together in a *coverage*. As its name implies, it contains the database schema. The items of interest in a *coverage* here refer to *features*. The libref *coverage* contains two type of *features*, that is, two *feature classes*. One is a *line feature class*, libref.lft; the other is a *text feature class*, libref.tft. Notice the .lft and .tft distinction in the file extensions. .lft designates a *line feature class*; .tft is for a *text feature class*. These are our spatial

		sample database		
		241.7 MB in disk		14 MB available
Name		Size	Kind	Last Modified
▽ 📁 my_db		—	folder	
▷	📁 my_lib	—	folder	
▷	📁 my_lib1	—	folder	
▷	📁 my_lib2	—	folder	
▷	📁 my_lib3	—	folder	
		—	folder	
▷	📄 dht			
▷	📄 dhx			
▷	📄 dqt			
▷	📄 dqx			
▷	📄 lat			
▷	📄 my_story.doc			
▷	📄 my_story.dox			

Figure 2.10　VPF database directory.

models. Every *feature*, of both classes, has associated attributes. *Line features* might have colors or line thickness. *Line features* might also represent reference features. In Figure 2.9, the library reference view shows continental outlines. Those are *line features*. Attributes are generally stored in the lft or tft table.

One set of attributes that are not stored in the *feature class* tables are the geometric attributes — that is, the coordinates that define the shape of continental outlines. Geometric attributes are stored separately in *primitives*. In Figure 2.12, there are *edge* and *text primitives*. The *edge primitive* table is edg, the *text table* is txt. The manner in which a *line feature* is constructed from a row in the *line feature class table* and one from the *edge primitive table* is described by an entry in the *feature class schema table*.

As introduced earlier, there are the edx and txx index tables in Figure 2.12. Another type of index table also shows up here: the *spatial index*. Spatial indices are designed especially for coordinate attributes. The *spatial index* for the *text primitives* is stored in the tsi *spatial index table*. The x index tables are mandatory. Spatial indices are optional, thus there might not be any esi for the *edge spatial index*. The *edge primitives* are required to have a *bounding rectangle table*, however. The *edge bounding rectangles* are stored in the ebr table.

Finally, there are two more tables shown in the figure. They are libref.ljt and libref.tjt. These jts are *join tables*. There are intermediate tables to handle different cardinalities in the *feature-primitive* joins.

Figure 2.13 shows the layout of a *tile reference coverage*. As noted above, the purpose of a *tile reference coverage* is to manage physical storage of data in a VPF *coverage*. This is a meta-data *coverage* and has certain content requirements specified in the Standard.

The tilref *coverage* is no different from the libref *coverage* in terms of structural organization. There are, of course, variations in the domain *feature classes* between the two. The aim in visiting this particular *coverage* is to introduce elements of a

		sample database		
		241.7 MB in disk		14 MB available
Name		Size	Kind	Last Modified
▽ 📁 my_db		—	folder	
▽ 📁 my_lib		—	folder	
▷ 📁 dq		—	folder	
▷ 📁 gazette		—	folder	
▷ 📁 libref		—	folder	
▷ 📁 my_cov		—	folder	
▷ 📁 my_cov1		—	folder	
▷ 📁 my_cov2		—	folder	
▷ 📁 my_cov3		—	folder	
▷ 📁 tileref		—	folder	
▷ 📄 cat				
▷ 📄 dqt				
▷ 📄 dqx				
▷ 📄 grt				
▷ 📄 lht				
▷ 📄 my_lib.dpt				
▷ 📄 my_lib.rpt				
▷ 📄 my_story.doc				
▷ 📄 my_story.dox				
▷ 📁 my_lib1		—	folder	
▷ 📁 my_lib2		—	folder	
▷ 📁 my_lib3		—	folder	
▷ 📄 dht				
▷ 📄 dhx				
▷ 📄 dqt				
▷ 📄 dqx				
▷ 📄 lat				
▷ 📄 my_story.doc				
▷ 📄 my_story.dox				

Figure 2.11   VPF library directory.

face topology implementation. Besides the *line* and text feature classes in the libref *coverage* presented in the last section, a tilref *coverage* includes *area features. Area features* employ *face primitives,* as *line features* use *edge primitives,* and *text features* with *tex primtivest.*

A number of the tables are now familiar. These include the tilref.tft *text feature class,* the txt *text primitive,* the tsi *spatial index table* and the txx *index table.*

A *feature class schema table,* fcs, still defines the schema for the *coverage.* There is also the *edge primitive* and its associated indices.

Among a few new tables is the fac, or *face primitive table.* A *face* is defined by a boundary. This boundary is made up of the now familiar *edges,* fortunately. Because *edges* now have additional responsibilities, in contrast to that in the libref *coverage,*

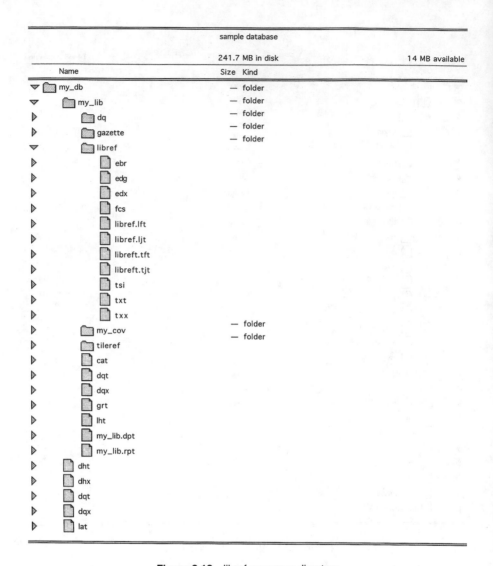

Name	Size	Kind
▽ 📁 my_db	—	folder
▽ 📁 my_lib	—	folder
▷ 📁 dq	—	folder
▷ 📁 gazette	—	folder
▽ 📁 libref	—	folder
▷ 📄 ebr		
▷ 📄 edg		
▷ 📄 edx		
▷ 📄 fcs		
▷ 📄 libref.lft		
▷ 📄 libref.ljt		
▷ 📄 libreft.tft		
▷ 📄 libreft.tjt		
▷ 📄 tsi		
▷ 📄 txt		
▷ 📄 txx		
▷ 📁 my_cov	—	folder
▷ 📁 tileref	—	folder
▷ 📄 cat		
▷ 📄 dqt		
▷ 📄 dqx		
▷ 📄 grt		
▷ 📄 lht		
▷ 📄 my_lib.dpt		
▷ 📄 my_lib.rpt		
▷ 📄 dht		
▷ 📄 dhx		
▷ 📄 dqt		
▷ 📄 dqx		
▷ 📄 lat		

sample database

241.7 MB in disk   14 MB available

Figure 2.12 libref coverage directory.

they must be ensured to connect properly. Otherwise, *face* boundaries might not close as expected.

Edge connections are maintained in the *connected nodes*, cnd, table. Connected *edges* form boundary *rings* around *faces*, resulting in the rng *ring table*. Fbr and ebr store *bounding rectangles* for *faces* and *edges,* respectively. Fsi and esi are the corresponding *spatial indices*.

This accounts for the 16 documents in tilref. Twelve of these 16 documents are needed to construct possibly the most simple *area feature class*. Fortunately, this is the most complex relationship in VPF, and all topological relationships, together

	Name	sample database				
		241.7 MB in disk				14 MB available
	Name	Size	Kind		Label	Last Modified
▽ 🗀 my_db		—	folder			
▽	🗀 my_lib	—	folder			
▷	🗀 dq	—	folder			
▷	🗀 gazette	—	folder			
▷	🗀 libref	—	folder			
▷	🗀 my_cov	—	folder			
▽	🗀 tileref	—	folder			
▷	📄 cnd					
▷	📄 csi					
▷	📄 ebr					
▷	📄 edg					
▷	📄 edx					
▷	📄 esi					
▷	📄 fac					
▷	📄 fbr					
▷	📄 fcs					
▷	📄 fsi					
▷	📄 rng					
▷	📄 tileref.aft					
▷	📄 tileref.tft					
▷	📄 tsi					
▷	📄 txt					
▷	📄 txx					
▷	📄 cat					
▷	📄 dqt					
▷	📄 dqx					
▷	📄 grt					
▷	📄 lht					
▷	📄 my_lib.dpt					
▷	📄 my_lib.rpt					

Figure 2.13   tilref coverage directory.

with their implementations, are predefined. To a database producer, the task is simply to focus on domain-specific features — *Surface drainage*, *Transportation*, and so on. Modelling these additional *features* does not require new structure or organization in VPF; the *primitives* are simply used or arranged differently.

Figure 2.14 shows a tile subdirectory in the my_cov *coverage*. Instead of finding the *primitive tables* directly in a *coverage* subdirectory as in the last two examples, a *tiled coverage* places all its *primitive tables* in a separate tile subdirectory. These tile subdirectories are defined in the *tile reference coverage*. Each *area feature* in the *tile reference coverage* represents one *tile* and points to one tile subdirectory.

Name	Size	Kind	Label	Last Modified
sample database				
241.7 MB in disk			14 MB available	
▽ my_db	—	folder		
▽ my_lib	—	folder		
▷ dq	—	folder		
▷ gazette	—	folder		
▷ libref	—	folder		
▽ my_cov	—	folder		
▽ my_tile	—	folder		
▷ cnd				
▷ cnx				
▷ csi				
▷ ebr				
▷ edg				
▷ edx				
▷ end				
▷ enx				
▷ esi				
▷ fac				
▷ fax				
▷ fbr				
▷ fsi				
▷ nsi				
▷ rng				
▷ tsi				
▷ txt				
▷ txx				
▷ tileref	—	folder		

Figure 2.14　A tile directory.

Do note that *feature class tables* and the *schema table* are not in the tile subdirectory. *Features* are not stored or partitioned by *tiles*, hence all *feature class tables* and the *schema table* are not shown in this figure.

Figure 2.14 shows 18 documents. Most of these are already familiar from the tilref discussion. There is one new entry. It is the *entity node primitive* and its end table. As *connected nodes* define connections or intersection points of *edges*, *entity nodes* are free floating points.

Finally, Figure 2.15 shows the documents contained in my_cov, the *coverage* directory. The listing is three pages long. This large number of documents might seem intimidating at first; however, they all do follow a relatively straightforward pattern. The most complex relationship was demonstrated in the *area feature* in the tilref *coverage*.

Let's take the *line feature class* again. The *line feature class* shown in Figure 2.15 is aline. The *feature class table* for this is aline.lft. The aline.lft table contains one

		sample database			
		241.7 MB in disk			14 MB available
Name		Size	Kind	Label	Last Modified
▽ 📁 my_db		—	folder		
▽ 📁 my_lib		—	folder		
▽ 📁 my_cov		—	folder		
▷ 📁 my_tile		—	folder		
▷ 📄 acmplxcl.cti					
▷ 📄 acomplex.cft					
▷ 📄 acomplex.cfx					
▷ 📄 acomplex.cjt					
▷ 📄 acomplex.cjx					
▷ 📄 acomplex.rjt					
▷ 📄 aline.lft					
▷ 📄 aline.lfx					
▷ 📄 aline.ljt					
▷ 📄 aline.ljx					
▷ 📄 alinecol.lti					
▷ 📄 anarea.aft					
▷ 📄 anarea.afx					
▷ 📄 anarea.ajt					
▷ 📄 anarea.ajx					
▷ 📄 anareacl.ati					
▷ 📄 anattri.rat					
▷ 📄 anattri.rax					
▷ 📄 apoint.pft					
▷ 📄 apoint.pfx					
▷ 📄 apoint.pjt					
▷ 📄 apoint.pjx					
▷ 📄 apointcl.pti					
▷ 📄 atext.tft					
▷ 📄 atext.tfx					
▷ 📄 atext.tjt					
▷ 📄 atext.tjx					
▷ 📄 atextcol.tti					

Figure 2.15   A coverage directory.

entry for each *feature* in this *feature class*. The *join table*, aline.ljt, is again, the join or key table that constructs a *line feature* by joining a row in the aline.lft table to one or more rows in an edg table or multiple edg tables from different *tiles*.

Whereas the *primitives* have spatial indices, *features* have *thematic indices*. Hence, alinecol.lti defines one *thematic index* on one particular column in the aline.lft table. There can indeed be multiple *thematic indices* stemming from one table. However, *thematic indices* in VPF can only be defined on single columns.

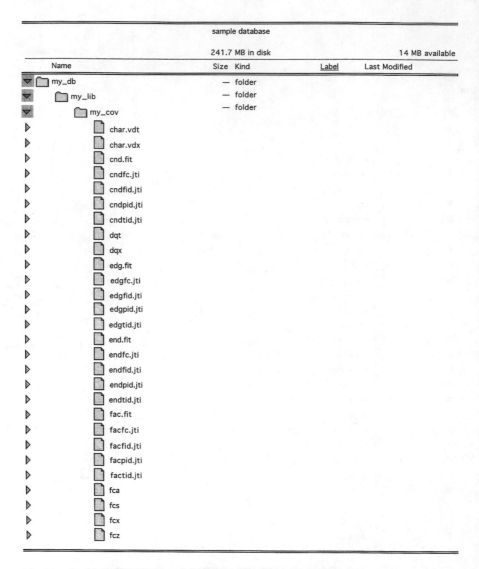

	sample database				
	241.7 MB in disk				14 MB available
Name		Size	Kind	Label	Last Modified
my_db		—	folder		
my_lib		—	folder		
my_cov		—	folder		
char.vdt					
char.vdx					
cnd.fit					
cndfc.jti					
cndfid.jti					
cndpid.jti					
cndtid.jti					
dqt					
dqx					
edg.fit					
edgfc.jti					
edgfid.jti					
edgpid.jti					
edgtid.jti					
end.fit					
endfc.jti					
endfid.jti					
endpid.jti					
endtid.jti					
fac.fit					
facfc.jti					
facfid.jti					
facpid.jti					
factid.jti					
fca					
fcs					
fcx					
fcz					

Figure 2.16   A coverage directory, cont'd.

This pattern is repeated five times. There is an area *feature class* and corresponding an area tables. The same is true for apoint, atext, and acomplex.

Acomplex implements a complex feature class. *Complex feature classes* are identical to regular, simple *feature classes*. Instead of referencing *primitives* for its geometric attribute as in simple *feature classes*, a *complex feature class* references other *features*. Other than this distinction, there is no difference between a *complex feature* and a *simple feature*. Since all join relationships are defined in the *feature class schema table*, no additional structure is required to accommodate this class of feaures.

Name	Size	Kind	Label	Last Modified

sample database

241.7 MB in disk   14 MB available

▽ 📁 my_db — folder
  ▽ 📁 my_lib — folder
    ▽ 📁 my_cov — folder

▷   📄 int.vdt
▷   📄 int.vdx
▷   📄 txt.fit
▷   📄 txtfc.jti
▷   📄 txtfid.jti
▷   📄 txtpid.jti
▷   📄 txttid.jti

Figure 2.17   A coverage directory, cont'd.

Another pattern is designed for indexing the *feature/primitive* joins. *Features* and *primitives* are always joined from different tables, whether or not a *join table* is involved. For example, the tilref features in Figure 2.13 are constructed without the use of a jo*in table*. It is because they are commonly one-to-one *feature*-to-*primitive* relationships.

However, this feature-to-primitive relationship can be indexed using the join index pattern. One join index table is constructed for each of the *primitive* types. Thus, the edg.fit is the *join index* for *edge primitives* and all *line* or *complex feature classes*. The *join index table* has four columns, and a *thematic index* is routinely built for each. Therefore, the join index pattern creates five tables for each *primitive* type. For the *edge primitives*, the *join index table* is edg.fit, and the four *thematic indices* are edgfc.jti, edgfid.jti, edgvid.jti, and edgtid.jti. Edg.fit is a name reserved in the Standard for the *edge join index table*. The four thematic indices can take on any name the database designer desires.

Therefore, even when at first glance a VPF *coverage* might contain a seemingly excessively large number of tables, these tables generally fall under one of these two patterns. A set of tables is constructed for each feature class. As additional *feature classes* are developed in the spatial model, a similar package of tables is reproduced.

CHAPTER 3

Data Structure

The data structure that underlies VPF is fairly simple. It is completely described by four elements; they are (1) tables, (2) data columns in the tables, (3) indices, and (4) directories.

In a VPF database all data are stored in VPF tables. Practically every piece of information a user wants to record in a VPF database is put into one of the tables. These are conventional tables of columns and rows. All these tables share one common table structure. One of the objectives of VPF is to have self-documenting databases. Tables are themselves self-documenting. The standard table structure is composed of a header section and a content section. The header in a table contains a schema that defines the columns in the table. Each column is assigned a name and has a data type. There is other information associated with a column definition. The complete list of data types is presented in Section 3.4.

Every record in a VPF table contains all the columns defined for that table. However, this does not mean that all records in a table are necessarily of the same length. There are a number of *string* data types in VPF that support variable width columns in a table. These *string* data types include the more familiar text or character strings. For geographic data, VPF implements a concept of coordinate strings as well. Instead of characters, coordinate strings are made up of coordinate values. Coordinate strings can be variable length arrays of coordinate pairs or triplets, representing two-dimensional coordinates and three-dimensional ones, respectively.

Besides tables and data columns, VPF also utilizes a number of index types. The purpose of these indices is solely to enhance the performance when accessing a VPF database. VPF indices are also very simple. None of the indexing schemes requires an extremely sophisticated navigation engine.

There are four types of indices in VPF. Among the four index types, the more familiar two are (1) the spatial index and (2) the thematic index. Spatial indices are defined on coordinate string columns, while thematic indices are defined on all other data types. Both thematic and spatial indices are built on a single column in one table. There is no multicolumn index required by the Standard.

The other two index types are a little obscure. These are (1) the variable length table index and (2) the join index. Any table in a VPF database that includes a variable length *string* data column or the *triplet id* data type automatically requires a variable length table index. The conceptual model for VPF is based on the notion of a georelational data model. Relations are constructed by joins on one or more base tables. All *features* in a VPF database are in fact relations built using one row in a *feature* table and at least one row in a *primitive* table. Join indices are principally defined to enhance the processing of these VPF *feature-primitive* relationships.

Indices, except for the join indices, have their own specialized peculiar structures. All of these index file structures are comparatively straightforward nonetheless. Join indices, on the other hand, are stored using the standard VPF table structure.

In a VPF database each table or index is stored in one separate file. These files are arranged, in a predetermined hierarchy, within directories. In Section 5.2.1.1 of the VPF Standard there is a description of a VPF directory structure. This is used where no native directory function is available. This structure is seldom seen in VPF database products. Practically all existing products depend on the host operating system for file management and directory supports. There were earlier discussions of 9-track tapes and other older operating systems that do not directly expose their directories to the end-users. However, this has not been a significant concern within the VPF user community.

Files in VPF are basically streams of bytes. In general, files and tables are equivalent in VPF. Individual tables are stored in and implemented by separated files. Unlike relational database management systems, there is no distinction between a table name and its corresponding file name. They are used almost interchangeably in most documents. For example, the *Library Header Tab*le, lht, introduced in the database tour is stored in a file named, incidentally, lht. The *inland water area feature class* shown in the sample VPF database is stored in an inwatera.aft table. The table is written in a file of the same name.

Records have no record separator in files. There is no column separator in records. The Standard does not dwell on files or directories. For the most part, files and directories are generally subsumed in the host operating system or in the storage media. Because there is not a separate implementation layer that removes tables from the physical files, table names are similarly restricted as are operating system files.

There are, nevertheless, residual concerns on directory path and file name references that map the host operating system file specifications to database items. Naming convention addresses one of these concerns.

NAMING CONVENTION

All file names and references to file names in a VPF database are required to be in lowercase ASCII characters. References to *database, library, coverage*, which are mainly implemented by directories instead of files, also follow the same naming convention. Similarly, column names and references to column names should also appear in lowercase ASCII characters.

This lowercase requirement has been introduced in the release of the Standard referenced in this book. Earlier specifications stated that names were not case-sensitive at all. For historical and typographical reasons, the Standard document shows upper case characters in its file name and directory name examples where, in fact, all file and directory names are to be in lowercase.

A number of file reference columns now allow non-ASCII character text strings as VPF continues to expand into other language encodings besides the North America Latin character set. The precise interpretations of those column definitions depend on whether the host operating system supports non-ASCII character file names. However, the conservative approach is to keep to the ASCII characters for the interim. Why should this cause concern? It lies in the fact that *feature class* names are used in table names and, ultimately, in physical file names.

File names are constrained by the ISO 9660 implementation for CD-ROM and earlier Windows/MS-DOS name length restrictions. All file names in VPF are required to observe the 8.3 convention. A file name, hence, is at most 8 characters long. The name can be followed by a period (.) and a file name extension that is made up of 3 or less characters. The first character of a file name must be an alpha character from "a" to "z". The remaining characters, up to seven, can be alphanumeric, including the underscore "_" character.

The file name extension, on the other hand, may contain only alpha characters from "a" to "z". There is, however, one more restriction on the choice of characters for the extension. The character "x" is not allowed in the last character in the extension for user-defined files. This is because variable length table indices generally adopt the name of the file for which the index is built. The variable length index file simply changes the last character in the file name to an "x".

This restriction on the last character of a file name extension should be restated. A user-defined table name, regardless whether an extension is used, and, hence, its consequent file name cannot end with the character "x". This is important because there are a large number of VPF files that do not use a file name extension. Many of these files do have variable length table indices.

Returning to the convention in choosing the last character of a table or file name, for a table or file named, e.g., myfile.ext, its variable length index file, if required, is called myfile.exx. This strategy works, but only to a limited extent, and database designers should be careful in their choice of file name. One can easily spot the problem if a database also contains another file named myfile.exb. Although myfile.ext and myfile.exb are easily distinguished from one another, their corresponding variable length indices cause a particularly tricky conflict.

In fact, such a conflict occurred in the Standard itself. In a *coverage* there are two tables describing various aspects of *feature classes* within the *coverage*. One is the *feature class attribute* (fca) *table*, the other is the *feature class schema* (fcs) *table*. The fcs is mandatory while the fca is only optional. Unfortunately, they cannot both follow the variable length index naming strategy by simply switching the last character in the file name to the character "x". Both of these would have used the identical file name. The index files would have written over each other. It was therefore decided that the variable length table index for the *feature class attribute table*, fca, will be fcx; for the *feature class schema table,* fcs, the index is named fcz. This is the only exception in which the name of a variable length table index does not end with the character "x". This exception is noted in the Standard. User defined tables cannot otherwise violate the variable-length table index convention.

RESERVED NAMES

The entire list of reserved file names is presented in three tables below. They are taken directly from the Standard. The three tables show (1) reserved file names, (2) reserved directory names, and (3) reserved file name extensions.

The use of reserved names represents simply a trade-off between theoretically possible unrestricted flexibility and pragmatic, operational considerations. There are the obvious advantages in maintaining a naming convention so that anyone handling a VPF database can easily understand its structure. Reserved names have various specific meanings in the VPF standard. A database designer is encouraged not only to observe the reserved names but also to enforce the conventions established by the Standards.

Table 3.1   Reserved Table Names

File Name	Description
cat	Coverage attribute table
cnd	Connected node primitive
csi	Connected node spatial index
dht	Database header table
dqt	Data quality table
ebr	Edge bounding rectangle
edg	Edge primitive
end	Entity node primitive
esi	Edge spatial index
fac	Face primitive
fbr	Face bounding rectangle
fca	Feature class attribute table
fcs	Feature class schema table
fsi	Face spatial index
grt	Geographic reference table
lat	Library attribute table
lht	Library header table
nsi	Entity node spatial index
rng	Ring table
txt	Text primitive
tsi	Text spatial index
char.vdt	Character value description table
int.vdt	Integer value description table

Table 3.2   Reserved Directory Names

Directory Name	Description
libref	Library reference coverage
dq	Data quality coverage
tileref	Tile reference coverage
gazette	Names reference coverage

Table 3.3   Reserved File Name Suffix or Extension

File Name Suffix	Description
.abr	Area bounding rectangle table
.aft	Area feature table
.ajt	Area join table
.ati	Area thematic index
.cbr	Complex bounding rectangle table
.cft	Complex feature table
.cjt	Complex join table
.cti	Complex thematic index
.doc	Narrative table
.dpt	Diagnostic point table
.fit	Feature index table
.frt[a]	Feature relations table[a]
.fti	Feature index table thematic index
.jti	Join thematic index
.lbr	Line bounding rectangle table
.lft	Line feature table
.ljt	Line join table
.lti	Line thematic index
.pbr	Point bounding rectangle table
.pft	Point feature table
.pjt	Point join table
.pti	Point thematic index
.rat	Related attribute table
.rpt	Registration point table
.tft	Text feature table
.tti	Text thematic index

[a] *Feature relations tables* are mentioned in DIGEST/VRF, and the frt extension is not strictly reserved in VPF.

TABLE STRUCTURE

All VPF tables share the same identical structure. Based on this structure a table is divided into three major parts. First there is a 4-byte count or offset. It gives the length of the header text and points to the beginning of the data content in the table. The 4 bytes represent a signed integer.

The length bytes are followed by a table description text. This second part of a table contains the table schema. These first two parts are generally referred to as the table header. This uniform header structure enables every table in a VPF product to contain its own schema, and it allows an application to process all tables consistently.

The last part of a table holds the data content. Every table must have all three parts. However, it is possible to have a table that is empty of content — that is, a table with all the definitions but with no data record. The content section is, thus, empty.

Normally, data values are stored in the content section according to the schema definitions. These values are repeated in rows of record. There is no record separator in VPF tables.

There are few restrictions in the makeup of tables in VPF. Each row in a table must have an id column. The values assigned to the id column start at one (1) for the first record encountered in a table and increase sequentially. Other than this row number column requirement, a table can contain anything that can be implemented using the VPF data types.

Table Header

Let's examine the structure a bit more carefully. A table begins with a length count of the header and a byte order code. The byte order code is required to support two of the more commonly used hardware architectures. This byte order code is optional. If it is not included, the table is assumed to use the least-significant-bit-first convention.

Table 3.4   Byte Order Code

Code	Description
M	Most-significant bit first
L	Least-significant bit first

Although it is possible to have different byte orders for different tables in one single database, it is seldom done. Indices do not have a byte order code in their headers. The byte ordering for binary encoding in an index follows that of the parent table. The byte order code also applies to the 4-byte count integer which precedes it.

Figure 3.1 shows the layout of the count bytes and the optional byte order code in a table header. The first four bytes of all VPF files, with the exception of some index tables, always contain a count of the header length. It is optionally followed by the byte order code. The count bytes and order code as terminated by a semicolon (;).

Count bytes and order code are followed by a table description. The description is also terminated by a semicolon. This description can be up to 80 characters long. The semicolon separator is not counted in the 80-character limit. Practically anything can appear within the description, but the description cannot include a semicolon.

However, if a description cannot be fully written within the 80-character limitation, a more verbose one can be stored separately in a narrative table. An optional narrative file name is the next item in the header. The file name, observing the 8.3

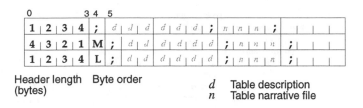

Figure 3.1   Count bytes.

convention, cannot exceed 12 characters in length. Once again, the narrative file name item in a table description is terminated by a semicolon.

The purpose of this description is to provide a more descriptive alias to the user of the database than the rather restrictive 8.3 table name. However, 80 characters are more than most applications can comfortably handle. Most VPF database viewers or browsers use the description in the user interface; an 80-character text does not really work well in drop-down lists or dialog boxes.

Figure 3.1 shows alternative ways to write the count bytes, the order code, and the table description. It is important to remember that the first four bytes represent a binary integer. Unlike the rest of the header, this integer is not written out as text characters.

Table Schema

The table schema is presented simply as a free formatted, delimited text string. A schema is a series of repeating column definitions. Although table schema are stored in text strings, the principal purposes of these descriptions are not formatted for reading by the end-user. White spaces or tabs are not necessary and generally not recommended to separate elements in the description text.

A table description contains one or more column definitions. Column definitions simply repeat in the table schema description text until all columns are properly described. Each column definition is separated by a colon (:). A semicolon terminates the list of column definitions.

Figure 3.2 shows the components of a column definition. The shaded characters in the figure are delimiter characters. Each column definition begins with a column name item. Column names are 16 characters or less. The column name item is terminated by an equals sign separator (=). In Figure 3.2, the column name is col.

A maximum of seven (7) definition attributes can be included in a column definition. Attributes are separated by commas in a column definition.

The first column attribute describes the data type. It is denoted by a single-character data type code. The type code shown in the sample in Figure 3.2 is I, designating an integer type. The data type code is followed by a comma (,) separator.

The next item in a column definition is the element count item. This means for a column of floating point numbers, for example, the column can contain an array of floating point numbers. The count item is an integer of no more than 3 digits.

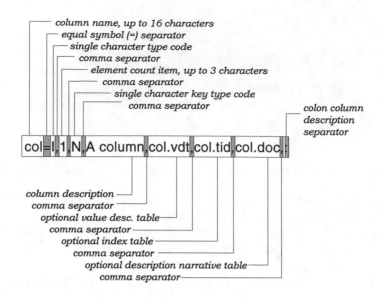

Figure 3.2   Column definition.

The count item, therefore, ranges from 1 to 999. It is 1 in the example, meaning the column contains a single integer value. In addition, an asterisk (*) can be used for variable length data columns — in this case the actual number of items in the field stored with the data, as a prefix to the array.

Although not specifically mentioned in the Standard, count values obviously should not be negative. Moreover, a count value of zero cannot be easily interpreted. A variable length column can conceivably be defined on any data type, but it is seldom done for types other than *string* data types. The count item, as all column definition items, is terminated by a comma separator.

After the count item, a key type code follows. There are three single-character key codes. An N is shown in the figure. This means the duplicated values are allowed in this column. A comma separator follows the single character code. The key type item is followed by a textual column description. This description can be of any length and is terminated by the comma separator. In many application programs written for VPF databases, this column description is commonly used to navigate the database. This serves the same function as that of the table description string. Care should be given in writing a concise description even though the field does not have a length limitation.

Three more file reference items complete the column definition. Since all three are file references, they are again constrained by the 8.3 naming convention and thus are 12 characters or less each. These tables are optional.

The first of these items references a value description table. This is a lookup table of sorts. Lookup tables are utilized, largely, in coded values columns. For example, a column of country codes might refer to a country code lookup table for an explanation or the complete country name based on a 2- or 3-character code.

The second references a thematic index table. A thematic index is simply an index defined on the column to enhance searches and queries.

The last of the three file references points to a narrative table for the column.

Each of these three items is terminated by a comma separator, as other items in the column definition. The entire column definition is itself terminated by a colon separator. There is no limit on the number of columns one can put into a table. Column definitions simply repeat. After all the columns have been exhausted, the table schema description ends. A final semicolon terminates the table schema description. This completes the entire table header.

Annotated Header Example

The following example is perhaps the simplest textual description of a table header possible. The count bytes and the byte order code are not shown. It is, nevertheless, a valid table header.

```
A table;tab.doc;col = I,1,N,A column,col.vdt,col.tid,col.doc,:;
```

This obviously does not describe a real table. There is only one column in it. In addition, that column is not the required row identifier column. This is translated into:

Table description is	"A table"
Table narrative file is	tab.doc
First column name is	"col"
Column type is	Long integer, type (I)
Number of items	1
Key type is	Nonunique
Column description is	"A column"
Value descriptions are in	col.vdt
Indexed by	col.tid
Column narrative file is	col.doc

Therefore, we have a table that is described simply as "A table." To find out more about this particular table, we have to read the narrative file called tab.doc.

This table contains only one column. The name of this column is "col." This column stores one single long integer per record. As declared in the key type attribute, the integer values in the table are not guaranteed to be unique. Thus, given any particular integer number as a search key, it is possible to retrieve more than one record from this table.

The description associated with the column is "A column." Additional description is hopefully available from the narrative file called col.doc.

The integer values stored in this column are actually coded values. Thus, to understand what those codes mean, a value description table, col.vdt, is referenced by the column. Moreover, this column is indexed, and the index is maintained in an index file calls col.tid.

The entire table description is 61 characters long. Adding to it the header length count integer of four bytes, the single character byte order, and the separator, the length of this table header is 67. This will also be the value assigned to the four-byte count at the very beginning of the table. The first position of a file that contains this table is assigned a position index of zero; the header length count integer value is found at this position. The data content section of the table begins at file position 67.

Because descriptions are not quoted, there are certain restrictions on what we can put in the table and column description fields. Considering the table description, since its separator is a semicolon, the description text itself, therefore, cannot contain a semicolon. A comma, even though it is a separator in column definitions, on the other hand, is acceptable here.

Hence, A table, a big table, is a valid table description. But, A column, a tall column, is not a valid column description, due to an embedded comma. A column description text cannot contain a comma because that is the separator in a column definition. The semicolon, on the other hand, is allowed in a column description text.

Second Annotated Example

This second table description example is included in the VPF Interface Standard document. We have copied it here to use as an example and will give some additional explanations. This example contains considerably more than the single column first example. In fact, this table does appear in a real product. Nevertheless, the interpretation of the table header is not much more complicated than the earlier example. Just a reminder here, the line separations shown in the figure are used purely for legibility considerations. The actual text description does not contain any line-feed or carriage return character.

Figure 3.3 shows the definition of what is actually a very simple table. The table contains 13 columns. The first column, required for VPF tables, is the row identifier column. It is essentially a sequence number. The remaining 12 columns are non-candidate key attribute columns. The id column is almost always treated as the primary key column in VPF tables.

Reading the definition of the first column, the id column, one notices that not all items of a column definition are included. The name of the column is id; it is of a long integer type. The column is identified as the primary key column for the table.

id = I,1,P,Row ID, — , — , — ,:

Since the id column is essentially a sequential number to identify the rows in a table, there is no need for the lookup table or an index. Dash symbols (–) are placed in these two items. The dash symbol (–) signifies that the column does not require a particular definition attribute. The column also does not have a column-specific narrative table. The last item in the column definition text is, therefore, again left with the dash symbol.

The second column, the f_code column, is the opposite of the first. Unlike the first, it utilizes every item in a column definition. The f_code column stores the

```
Surface Drainage Area;
src.doc;
id=I,1,P,Row ID,-,-,-,:
f_code=T,5,N,FACS Code,char.vdt,f_code.ati,f_code.doc,:
rgc=I,1,N,Railroad Gauge Category,int.vdt,-,-,:
hyc=I,1,N,Hydrographic Category,int.vdt,-,-,:
hfc=I,1,N,Hydrographic Form Category,int.vdt,-,-,:
exs=I,1,N,Existence Category,int.vdt,-,-,:
wid=F,1,N,Width (meters) category,-,-,method.doc,:
wvl=I,1,N,Water Velocity Average (m/sec),int.vdt,-,-,:
wda=I,1,N,Water Depth Average (meters),int.vdt,-,-,:
mcp=I,1,N,Material Composition Primary,int.vdt,-,-,:
dvr=I,1,N,Dense Bank Vegetation Right,int.vdt,-,-,:
dvl=I,1,N,Dense Bank Vegetation Left,int.vdt,-,-,:
bgr=I,1,N,Bank Gradient (Slope) Category Right Bank (%),-,-,:;
```

Figure 3.3   Schema example.

5-character FACS codes which describe particular physical features on the ground. It is a fairly comprehensive coding scheme for geographic features and phenomena. FACS codes have since been obsoleted and replaced by a different coding system. However, this is beyond the scope of the discussion here.

```
f_code = T,5,N,FACS Code,char.vdt,f_code.ati,f_code.doc,:
```

The code is obviously not unique in a database because multiple features of the same type are quite likely. The f_code column constitutes a likely access path to the data, therefore an index is included. The thematic index table defined for this column is indicated in the f_code.ati entry. The name f_code probably is not very meaningful, hence the slightly more descriptive FACS code text is included. Even with this expanded alias, additional explanations might be warranted to give the database users a better understanding of what the FACS coding scheme is, thus a narrative table is included and referred by f_code.doc. The lookup table is called char.vdt.

The remaining column definitions can be read in exactly the same manner. None of them presents any surprise. An int.vdt lookup table appears in a few definitions in this example. This is an example of how one single lookup table can be reused by multiple columns, and perhaps, by many tables. For example, in some database, the value of -9999 is used to identify *unknown* in all measurement schemes. Hence, more than one measurement attribute column can possibly use the same lookup table.

Third Annotated Example

The final example of a table description text shows a short-hand representation of the text. In a table description, VPF allows a designer to exclude trailing unused items in a column definition. The table description shown in Figure 3.4 defines a table identical to the one presented earlier.

Hence, if the last three items of the column definition are not used in the id column, they can be safely removed from the description text, without even the

```
Surface Drainage Area;
src.doc;
id=I,1,P,Row ID,:
f_code=T,5,N,FACS Code,char.vdt,f_code.ati,f_code.doc,:
rgc=I,1,N,Railroad Gauge Category,int.vdt,:
hyc=I,1,N,Hydrographic Category,int.vdt,:
hfc=I,1,N,Hydrographic Form Category,int.vdt,:
exs=I,1,N,Existence Category,int.vdt,:
wid=F,1,N,Width (meters) category,-,-,method.doc,:
wv1=I,1,N,Water Velocity Average (m/sec),int.vdt,:
wda=I,1,N,Water Depth Average (meters),int.vdt,:
mcp=I,1,N,Material Composition Primary,int.vdt,:
dvr=I,1,N,Dense Bank Vegetation Right,int.vdt,:
dvl=I,1,N,Dense Bank Vegetation Left,int.vdt,:
bgr=I,1,N,Bank Gradient (Slope) Category Right Bank (%),:;
```

Figure 3.4 Schema example.

```
Surface Drainage Area;
src.doc;
id=I,1,P,Row ID,:
f_code=T,5,N,FACS Code,char.vdt,f_code.ati,f_code.doc,:
rgc=I,1,N,Railroad Gauge Category,int.vdt,:
hyc=I,1,N,Hydrographic Category,int.vdt,:
hfc=I,1,N,Hydrographic Form Category,int.vdt,:
exs=I,1,N,Existence Category,int.vdt,:
wid=F,1,N,Width (meters) category,-,-,method.doc,:
wv1=I,1,N,Water Velocity Average (m/sec),int.vdt,:
wda=I,1,N,Water Depth Average (meters),int.vdt,:
mcp=I,1,N,Material Composition Primary,int.vdt,:
dvr=I,1,N,Dense Bank Vegetation Right,int.vdt,:
dvl=I,1,N,Dense Bank Vegetation Left,int.vdt,:
bgr=I,1,N,Bank Gradient (Slope) Category Right Bank (%),:;
```

Figure 3.5 Schema example.

inclusion of a dash symbol (–). However, the dashes are required in the wid column because the unused items appear in the middle of the description (Figure 3.5). This form of description is not very common in databases being produced at this time. Most product specifications do include the dash symbol to explicitly indicate that a column definition attribute is not used.

DATA TYPES

For the most part, data types available in VPF are strictly conventional. Most of the data types follow well-established standards elsewhere.

In this section, we focus on three specific topics about some of the not so conventional data types. These include *string* data types, which include also variable length *strings*; and the triplet Id data type. The last subject deals with the Null data

Table 3.5   Data Types

Type Code	Definition	Descriptions
T	Text strings	ASCII, ISO 2022
L	Latin-1 text strings	ISO 8859
N	Full Latin text string	ISO 6937
M	Multilingual text string	ISO 10646
F	Short floating point number	ANSI/IEEE 754, 32-bit
R	Long floating point number	ANSI/IEEE 754, 64-bit
S	Short integer number	
I	Long integer number	
C	Short 2-D coordinate string	
B	Long 2-D coordinate string	
Z	Short 3-D coordinate string	
Y	Long 3-D coordinate string	
D	Data and time data type	ISO 8601
X	Null data type	Length = 0
K	Triplet Id data type	

type in VPF. The earlier examples in this chapter already introduced data types "I", "F" and "T". Table 3.5 contains the complete list of allowable data type in VPF.

String Data Type

The *string* data type was introduced earlier in this chapter. Moreover, the element count column attribute is most commonly used with *string* data types. For example, in Figure 3.3, the T,5 in the f_code column denotes a 5-character-wide text string column.

For fixed-length string data types, such as the example above, the number of repeating elements in a string is defined in the table schema. This is fairly obvious. The count item contains a maximum of 3 digits; hence, a fixed-length string cannot exceed 999 elements.

Wherever a variable length column is defined, an asterisk is entered in the element count item in the table schema. Therefore, to indicate a variable length text field, rather than T,5, a T,* is defined instead. In this instance, a 4-byte integer count is prefixed to the string that is actually stored in the field. This count behaves in exactly the same manner as the item count definition for fixed-length strings, except the maximum is limited by what can be represented in a 4-byte integer. Again, this prefix count value is a signed integer. VPF does not have an unsigned number type.

Outside of VPF, the *string* data type is commonly encountered as character strings. The applications in VPF for these *string* data types are not very different from, e.g., the CHARACTER and CHARACTER VARYING (or VARCHAR) in SQL. However, since VPF is designed specifically for spatial, or geographic data, this notion of *string* is also extended to the spatial domain.

Coordinate strings are arrays of elements called *coordinate data types*, which are themselves a strange breed; this is because a *coordinate data type* is itself a 2- or 3-element array. In VPF, positions are measured either in 2-dimensional or in 3-dimensional Cartesian coordinate space. Each coordinate data type in a

2-dimensional space is represented by two floating point numbers in VPF. This is called a coordinate-pair. Each coordinate data type in a 3-dimensional space is represented by three floating point numbers. The 3-dimensional coordinate object is a coordinate triplet. In both types, the x coordinate is always listed first, the y coordinate second and, in the triplet, the z coordinate third.

A coordinate string is simply an array of coordinate pairs or coordinate triplets. There are four coordinate string types: C, B, Z, and Y. A "C" data type is a 2-dimensional coordinate string using short floating point numbers; a B data type uses long floating point numbers. A "Z" data type is a coordinate triplet that uses short floating point numbers; and a "Y" type is one that uses long floating point numbers.

Similar to text or character strings, coordinate strings can be either of fixed length or of variable lengths. Again, the length of a coordinate string is stored in the column definition for a fixed-length string, or prefixed to the string itself if it is of variable length.

Conceptually any data type in VPF, such as an integer, is allowed to have repeating values in one single column. This is declared simply by using the same count item in the schema definition as in the string discussions. However, a count value is seldom set to have a value greater than one for types other than the text and coordinate strings.

Triplet Id Data Type

The second data type discussed here is possibly the most unpopular data type in VPF; it is the *triplet id*. A *triplet id* column is used solely as a foreign key to access rows in graphic *primitive* tables. More specifically, these are mostly primitive tables in *tiled coverages*. In the most fundamental sense, the *triplet id* data type allows compound keys to be defined in a VPF database.

A *triplet id* column replaces the more mundane integer foreign key needed to reference a graphic in a *tiled coverage*. Graphics are always identified by their row number in graphic primitive tables. Row numbers are essentially used as primary keys in these tables. A *coverage* is a logical dataset. A *tiled coverage* subdivides this logical construct into storage subsets but all elements remain in a logical dataset. In this *tiled coverage*, graphics thus cannot be uniquely identified by their individual sequence numbers alone. Their individual row identifiers must be qualified by an identifier from the *tile* in which they belong.

Since we have not discussed the concept of tiling, let's use a simple example to illustrate the use of this *triplet id* in a VPF database. Suppose we have a set of randomly placed points within a square. For each of these points we identify a neighbor that is nearest to it. To record this relationship in our database, we can simply add an additional column to the table that was originally designed for our random points. To maintain the nearest neighbor relationships in the table, names of the nearest neighbor point are put into the new column.

If, for whatever reason, the original set of points is divided into two sets depending on whether a point falls in the left half or the right half of the square, the majority

Table 3.6   Triplet Id Field Codes

2-bit Type Code	Descriptions
0	Length = 0
1	Length = 8 bits
2	Length = 16 bits
3	Length = 32 bits

of the nearest neighbor pairs would still remain entirely within one set or the other. Some pairs, perhaps a very small number, might be separated from each other as they are located in different halves, or *tiles*. This division process is essentially tiling. It is for these separated pairs that the triplet id data type is created.

To handle these split relationships, whenever an object needs to reference another object that is not found within the same set, a set, or tile, qualifier would be included.

This new information can be straightforwardly maintained by the addition of yet another column to hold the qualifier without the need to create a new data type. However, most of the entries in this new column are likely to be empty because very few of the nearest neighbor relationships would actually involve one point from one set and another from a different set. The triplet id was thus invented to minimize these wasted spaces — unfortunately, at a not inconsiderable cost.

A *triplet id* data type is of variable length. Including this data type in a table automatically causes the creation of a variable-length table index, while the element count in the column definition could still show a value of one. The actual byte length of a triplet id is determined by an 8-bit type field at the beginning of the data field. This 8-bit type field is separated into four 2-bit type subfields. Each of these subfields defines the length of the actual data content length for one component of the triplet id. The subfield length code is shown in Table 3.6.

Since there are four such subfields, we can actually have four components in a triplet id data type. Alas, the fourth subfield is not used. So it is still true that this is a triplet for now.

In this scheme, the first component in the triplet id data type stores the internal integer foreign key, i.e., ones that do not require a tile qualifier. The second component stores the tile qualifier, and the third stores the external foreign key, for those that do require a qualifier. The examples in Figure 3.6 illustrate the usages of the triplet id data type.

The first example in Figure 3.6 references a single-byte internal identifier. That is where the element that uses the reference and the element being referred to are both located in the same *tile*. No tile number qualifier is needed.

The second and third examples are essentially the same as the first, except that the second example shows a reference using a short, 2-byte integer, while the third is a long, 4-byte integer.

The fourth example shows where a compound key that usually takes up two columns can be combined into a single column *triplet id* data type column. The internal foreign key component of the triplet is unused.

The usage presented in the last example is encountered almost exclusively in the *primitive* tables to maintain cross-tile topology. Here, the *graphic primitive* in

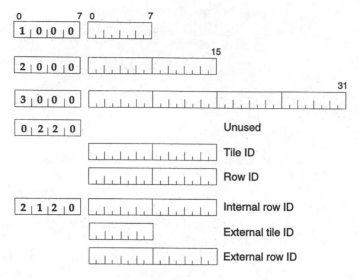

Figure 3.6   Triplet Id.

question has both a relationship with a neighbor that is found within the same *tile* and another with a neighbor that falls on an adjacent *tile*.

Null Data Type

While the *triplet id* data type is the most unpopular, the *Null* data type introduced here is possibly the most ignored data type in VPF.

First, the *Null* data type in VPF is completely different from null values that we can assign to other data types. A null text string cannot be represented by a *Null* data type. A null text is an empty string or one with a width of zero. Similarly, a *Null* data type cannot store a null integer.

Why do we need a *Null* data type? Based on the initial architecture of the VPF data model, a *Null* data type is compatible with any data type. The *Null* data type allows tables that contain a column of such *Null* data to be merged, unioned, with another table, if every other non-*Null,* column matches up between the two tables.

There are currently only two such columns mentioned in the Standard. They appear in the *connected* and *entity node* tables. These two node tables were initially one single table. It was somehow separated during the early development phases of the VPF Standard. Thus, to ensure that the tables can indeed be put back together, the *Null* data columns are inserted.

Therefore, the containing face column in the connected node table is marked as a *Null* data column. Similarly, the first edge column in the entity node table is a *Null* data column. We'll return to these tables later on, so it will suffice for now to point out that a containing face cannot possibly be defined for a *connected node* in VPF, and an *entity node* can never have a first edge.

These two *Null* columns are in fact optional columns. Even though they do show up in the Standard document, they are not required to be included in every VPF database. Seldom are there reasons to justify keeping a *Null* data column in these or other tables. However, a *Null* data column has zero width. It does not consume any space in the data content section.

CHAPTER 4

Features and Feature Classes

It is fundamentally important to remember that *features* are generally the only things an end-user should be interested in when working with a VPF database. A large part of what we have introduced in the earlier chapters and certainly all we'll discuss later on should be transparent to database users.

Features are the logical elements to model a geography. *Features* have attributes that describe their properties. In the VPF view of the world, one of those properties a *feature* has is its geometry. The geometric property of a *feature* is described using *primitives*. The geometric property of a river *feature* might be the meandering lines marking its banks. The footprint of a building *feature* is a geometric property. Every VPF *feature* must have at least one *primitive*. Most *features*, indeed, have more than one.

Features are grouped into *feature classes*. Classes are collections of *features* that share a common set of attributes and are made up of similar *primitive* types. In a database of political entities, a *feature class* might be defined for the countries. Each individual country is, in turn, one *feature* of the *country feature class*.

There is one *feature class* type for each of the corresponding *geometric primitive* types. Briefly *geometric primitives* are the conventional zero-dimensional geometric points, one-dimensional lines and two-dimensional areas. Although VPF supports Three-dimensional coordinates, geometric volumes are not implemented in VPF.

An additional *primitive*, the *text primitive*, is specifically designed to support cartographic applications. *Text primitives* allow the designer to place text labels anywhere on the map to record geographic phenomena that cannot otherwise be conveniently depicted by other geometric *primitives*.

A point type *feature class* corresponds to zero-dimensional geometry, a line type to one-dimensional geometry, and an area type to two-dimensional ones. There is also the text *feature* type for the *text primitive* as well.

However, one seldom finds a generic line or point *feature class* in a VPF database. The modeller almost always implements these as, e.g., the class of *highway features* or *well features*. Therefore, from a single collection of one-dimensional geometries, we can construct a comprehensive transportation network containing an *interstate highway feature class*, an *artery road feature class*, a *local street feature class*, a connector or *frontage road feature class* and, perhaps, the *ramps and interchange feature class.*

The separation of *geometric primitives* and modelled domain *features,* together with the application of relational modelling, provides tremendous flexibility in modelling a geography using VPF. A single geometric piece, for example, a line segment, might be used in multiple *feature classes.* Sections of an artery might belong to the interstate highway network. Since they both share a common geometric *primitive,* there is no problem with aligning two disparate graphics in the database. Conversely, an individual *feature* might be built from multiple geometric pieces. "Interstate 95" might include any number of line segments.

Furthermore, not all *feature classes* are restricted to using a single dimensional *primitive* type. *Complex feature classes,* as their names suggest, are made up of other *feature classes* and, possibly, complex geometries. A *feature class* of the complex type can be built to model, for example, surface hydrographic systems. Each *feature* in this class could contain any combinations of streams, lakes, perhaps waterfalls, and so on.

COVERAGES AND TILES ORGANIZATION

Feature classes and *features* are defined for *coverages.* In terms of a VPF database, *features* from different *coverages* are essentially independent of each other. The Standard does not include any mechanism to support logical relationships across different *coverages. Features* from multiple different *coverages,* however, do have a spatial relationship because all *coverages* in a library share the identical coordinate reference system.

In DIGEST/VRF, relationships among features from different coverages are possible. These relationships are implemented in a feature relations table. A relationship between two features might take the form of:

database1\library1\coverage1\onefeature.fft   relates to

database1\library1\coverage2\anotherfeature.fft.

Since VPF does not yet have *features* that cross *coverages,* it is therefore important to keep this restriction in mind when organizing data in a database into *coverages.* Although the *feature relations table* in DIGEST provides a means to explicitly record relationships among *features* in different *coverages,* this approach still does not allow a *feature* to use the *geometric primitives* from any coverage other than its own. Returning to the hydrographic system example above, a designer must ensure that all component *features* of the *complex surface hydrography feature class* are contained within the same *coverage.* Hence, unless both the *stream feature class* and the *lake feature class* are maintained in the same *coverage,* it is not possible to define a *complex hydrography system feature* that is composed of *stream* and *lake features.*

Having made the observation that there is no cross-coverage feature class, we must reiterate an earlier comment that the VPF Standard specifies only the minimum foundation to model a geography. The Standard defines a set of data structures–building blocks (Chapter 3) and a set of relationships. The underlying model in VPF is a relational table. If a VPF product adds a table at the library level that references cross-coverage relationships and satisfies a particular application requirement, it is fully entitled to use such implementation and still remains absolutely compliant to the Standard.

Finally, regarding *feature class* organizations and *tiling*, on the other hand, *features* and *feature classes* are unaffected by a tiling scheme. VPF does include a mechanism to support spatial and logical relationships among *features* across different *tiles* in a *coverage*.

To the database end-users, tiling should be entirely transparent. Notwithstanding performance considerations, since tiling is fundamentally a performance-enhancing technique, any operation on a database should yield the identical result whether or not tiling is applied to the database, and altering the tiling scheme of a database should not affect the result of an operation either.

CATEGORIES OF FEATURES

Feature classes can be divided into categories. There is the category of *simple feature classes*, which is roughly interpreted as *feature classes* that are made up of one *primitive* type only. There is also the category of *complex feature classes*. *Complex feature classes* are made up of other *feature classes*. All *features* must ultimately reference some *primitives* in the database. *Simple features* do this directly, while *complex features* reference *primitives* indirectly, through other *features* that they reference.

The category of *simple feature* types is further subdivided into four different subtypes. Each of these subtypes represents one geometric type. The simple feature types are: *area features, linear features,* and *point features*. In addition, there are *text features*, which corresponds to *text primitives*. A *complex feature class*, on the other hand, might involve one or more geometric types.

Area Feature Class

Area feature classes represent two-dimensional, polygonal objects in a spatial model. For example, soil polygons are *area features*. The class of soil polygons contains numerous areal features of different soil properties — soil group, texture, depth, drainage, and so on. Other *area feature classes* include parcels, city blocks, and service territories. Countries, continents, and lakes are also example of *area feature classes*.

Features in an *area feature class* can overlap. For example, service territories may overlap so that a location might have its redundant service coverages from two regional offices.

Areal features can also be composed of multiple, discontinuous components. For example, a country may occupy a mainland and multiple offshore islands. The *country area feature* is then an aggregate of all these discontinuous polygons.

Line Feature Class

Familiar *line features* include highways, powerlines, and pipelines. Boundaries of *area features* may be members of a *line feature class*. For example, the border of a country can be represented by a *line feature*. Similar to *areal features*, *line features* may overlap. The example of one road segment being given both an inter-state highway and a local street designation is a situation where *line features* overlap.

Line features can be directed or undirected. A directed *line feature* is used for indicating a direction of travel along the *feature*. A *line feature* that represents a one-way street might need to be a directed line feature to properly model the traffic flow along that street. A stream might also require a directed *line feature*.

Point Feature Class

Although mentioned earlier that we seldom encounter *point features* in real life, these *features* are quite common in a spatial database. On an aeronautic navigation database, airports might be represented in a *point feature class*. After the plane lands and begins taxiing, that *point feature* might have to take on another representation.

The *point feature class* is used to model free-standing objects. These may include airports, buildings, buoys in the ocean, or moving ships. *Point features* also represent attached objects such as pumping stations along pipelines, road intersections, and entrance or egress points of a park.

Text Feature Class

Text feature classes are basically escape routes for geographic features that do not have a well-demarcated boundary. Sometimes this is referred to as a *cartographic feature* rather than the other *geographic features*. It is because even though these geographic objects or phenomena are not easily delineated in real life, they are drawn on maps. Since they are common in cartographic representations, it is therefore necessary for the database to properly maintain them.

Text feature classes are separated from other types because it is difficult to perform analytical operations on them precisely. Unlike the other types, there are also few rules to govern their placement. Examples of these *features* are: the Pacific Ocean, Silicon Valley, or the Rockies. Depending on the particular application, the precise boundaries of these objects may not be relevant. However, their relative position might provide very useful locational references.

Complex Feature Class

A *complex feature class* is a *feature class* that is made up of one or more other *feature classes*. The underlying geometric dimensions of the *primitives* are no longer particularly relevant at the *complex feature class* level. One *complex feature class* can be composed of two *point feature classes*, one *line feature* class and an areal one, or some other combination that also includes a *text feature class*.

An obvious example of a *complex feature class* is one that models a hydrographic system. A particular *feature* in this class might include a stream that feeds into a lake. It is common practice that, at some scale of representation, a stream becomes so narrow that its water surface collapses into a one-dimensional line rather than the more natural two-dimensional surface.

Another not so obvious use of a *complex feature class* reflects an idiosyncrasy of *point features* in VPF. Take the example of representing the different sources of water in a VPF database. We might decide to use a single *point feature class* to represent these sources.

A *feature class* might be built using free-standing point objects. These might be artesian wells. On the other hand, there might be pumping stations or outlets from aqueducts or pipelines that supply water. These second kinds of outlets are not free-standing objects; instead, they are attached to other, perhaps linear, objects. A *simple point feature* class cannot be defined using both free-standing point objects and attached point objects because they belong to different *primitive* types (Chapter 5). Therefore, it may be necessary to define a *wells point feature class* and another, e.g., *pumps feature class*. The water source feature class thus becomes a *complex feature class* utilizing these two component *point feature classes*.

FEATURE CLASS IMPLEMENTATION

The implementation of *feature classes* and *features* in VPF is entirely flexible. As stated earlier, the Standard does not have predefined *feature classes*, with two exceptions: *data quality*, dq, *feature class* (Chapter 10) and the *tile*, tileref, *feature class* (Chapter 9). Besides these two, all *feature classes* in a VPF database are completely determined by the application domain rather than the Standard.

Each *feature class* has one base feature table. *Features* belonging to the *feature class* are represented by records in the feature table. Some, or all, of the properties related to a *feature class* may be stored on the base feature table. A database developer is allowed to put any number of columns into the table. Properties may also be stored in other related tables. This is governed solely by the database design objectives for one specific product.

The base table has two requirements. As in all VPF tables, each record has an id column that uniquely identifies a *feature*. Second, one of these user-defined attributes must be a reference to either *features* in another *feature class* or *primitives* that makes up the geometry part of the *feature*. Geometric descriptions of *features*, by and large, are not stored in the feature class tables directly.

In sum, there must be as many feature class tables as the number of *feature classes* defined for a *coverage*. Dependent upon the approach used in its implementation, each of the feature class tables may or may not have a companion *feature join table*. In some databases one might also encounter a number of *feature join index tables*.

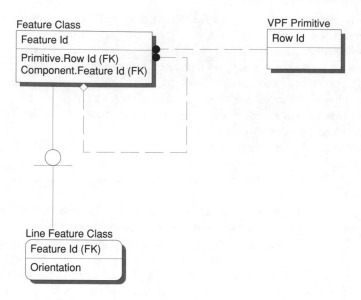

Figure 4.1   Feature/primitive join.

These two kinds of additional tables often appear in a VPF database. The *feature join tables* handle one-to-many relational joins. *Join index tables* are performance enhancers. Neither of these affect a geographical model substantively.

Figures 4.1 and 4.2 depict the relationships among the tables presented here. These figures show two different implementation approaches. Both approaches are available to the developer. In fact, it is common to see both approaches in a single database. The reason that they are separated in our diagrams is mainly for clarity and legibility.

Figure 4.1 shows three tables. All three are essentially templates — that is, we do not really have a table called feature class. We have instead country feature class table, highway feature class table, island feature class table, and so on. All these — country, highway, island feature class — tables are instances of the feature class table template.

Figure 4.1 illustrates a *feature/primitive* implementation using a primitive pointer on the feature table. This method is only applicable where every *feature* in the *feature class* uses one and only one *primitive*; there is no assumption that all *primitives* are used in *feature* definition. The figure also shows one special subclass of the feature table. The *line feature class* sometimes contains an orientation property which allows the implementation of directed *line features*. A positive value, usually a value of +1, indicates that the line feature follows the same direction of travel as that of the primitive it employs. A negative means that the two travel in opposite directions.

Figure 4.2 shows a slightly different implementation of a similar relationship. In this case the pointer goes onto the primitive table. This is only possible when every *primitive* is used by at most one *feature*. Each *feature* in the *feature class*, on the other hand, may utilize multiple *primitives*. Furthermore, some *primitives* might not be used at all.

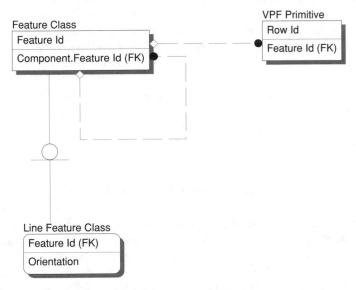

Figure 4.2   Feature/primitive join.

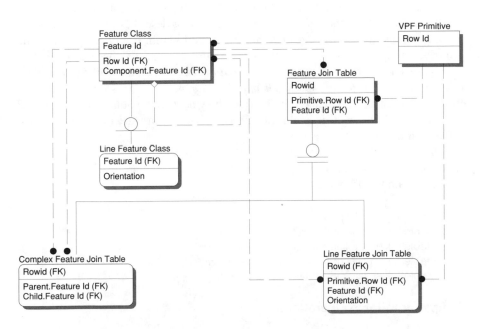

Figure 4.3   Feature/primitive join.

Figure 4.3 further extends the same relationships as those depicted earlier. Where a feature in the feature class might reference more than one primitive and a primitive might be simultaneously used in more than one feature, the many-to-many relationships is decomposed into the feature join table. Join tables are fairly common, even in situations that do not have a many-to-many relationship between features and primitives.

Where a line join table is used, the orientation column is relocated from the line feature table to the join table.

Finally, it is not uncommon to have all three different strategies applied in the same dataset.

FEATURE CLASS SCHEMA TABLE

All these relationships must be described somehow. The *feature class schema table* completely describes all relationships in a *coverage*. There will always be one *feature class schema* (FCS) table for a *coverage*. This table describes how the *feature classes* are put together. Optionally, some *coverages* would also include a *feature class attribute* (FCA) table which lists all *feature classes* defined for the *coverage*.

The *feature class attribute* (FCA) table lists all *feature classes* in a *coverage* rather than attributes of features. The table has three columns: (1) fclass, for feature class name, (2) type, for *feature class* type, and (3) descr, for a short text description of the *feature class*. The name for *feature class attribute table* is taken from the convention that has been established for all of the major organizational elements in a VPF database. There is a *library attribute table* that describes all the *libraries* in a *database*, and a *coverage attribute table* is found within each *library*. Thus, the *feature class attribute table* enumerates all *feature classes* in a *coverage*.

Figure 4.4 shows the content of the *feature class attribute* and *feature class schema* tables. The *feature class schema* (FCS) table is perhaps one of the more critical tables in a VPF database. It describes how data are combined to form a VPF data product. Without it, it is almost impossible to reconstruct even the simplest *features* in a VPF database.

The *feature class schema table* itself is relatively straightforward, however. The table has the mandatory record identifier, id, column. There are five columns other than the identifier. They are:

1. feature_class, the *feature class* name
2. table1, the name of the first table of a join
3. table1_key, the join attribute in the outer table
4. table2, the name of the second table of a join
5. table2_key, the join attribute in the inner table

Sample Join Table Operations

Although the *feature class schema table* might seem confusing at first, it basically maintains very simple relational joins. Using these queries, a complete representation

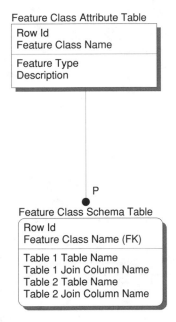

Feature Class Attribute Table

| Row Id |
Feature Class Name
Feature Type
Description

P

Feature Class Schema Table

| Row Id |
Feature Class Name (FK)
Table 1 Table Name
Table 1 Join Column Name
Table 2 Table Name
Table 2 Join Column Name

Figure 4.4   Feature class schema table.

of geographic objects can be easily constructed from a *feature class* table, a *primitive* table and any number of other related-attributes tables.

Let's take, for example, a street *feature class*. These are regular streets in an everyday city. Since we are taking a very generalized view of the geography, we model the streets by their centerlines rather than by detailing pavements and gutters. In our database, some streets have alternate names in addition to their standard street names.

Figure 4.5 shows the relevant tables in a database. The first table is the mystreet.lft table. It contains three columns. The first column in mystreet.lft is the standard VPF record identifier column id. The other columns join key columns so that we can combine data from the other tables to construct a representation of *street features*. The edg_id is a foreign key to the edg *primitive* table. The edg table is used to maintain the geometric *edge primitives*; the edge shape is stored in the coordinate column.

The name column allows us to perform a relational join with the mystreet.rat related-attributes table. It is from this table that we retrieve the alternate names.

The *feature class schema table,* therefore, has entries to represent the two join operations.

One obvious peculiarity to note from this example is that records are repeated. The first two rows in Table 4.1 are essentially identical for all practical purposes, especially in the relational sense. The same is true for the third and fourth rows.

This duplication is purely the result of an implementation consideration during the original drafting of the Standard. Relational joins, especially those equi-joins, are by definition both commutative and associative. VPF applications are expected

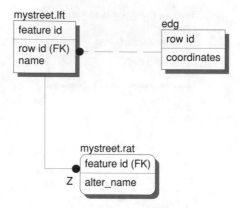

Figure 4.5 Street feature class example.

Table 4.1 Sample Feature Class Schema Table

feature_class	table1	table1_key	table2	table2_key
mystreet	mystreet.lft	edg_id	edg	id
mystreet	edg	id	mystreet.lft	edg_id
mystreet	mystreet.lft	name	mystreet.rat	name
mystreet	mystreet.rat	name	mystreet.lft	name

to take advantage of this and of other characteristics of the relational data model, together with information about a VPF database, to deduce the optimal access path when manipulating the tables in a database.

However, elaborate query optimization routines are not always cost effective, nor are they necessary in every VPF application. To enable applications without a query optimization module to manipulate the tables, the definitions in the FCS explicitly enumerate the various access paths.

Having just made the point about access paths, it is important to note that, unfortunately, the ordering of definitions in the FCS ensures neither the efficiency nor the effectiveness of an access path. The FCS table merely defines how relationships among the tables can be reconstructed. It remains for the client application of a VPF to develop an efficient navigation strategy.

In general, the four entries in the FCS table above roughly represent the following SQL. Although the feature class schema does not explicitly indicate the type of joins in its entries, certain assumptions are implicitly specified. Because the Standard states that every *feature* must have at least one *primitive*, the join between the feature class table and the primitive table is a natural equi-join. It cannot be an outer join because not all *primitives* are referenced in any particular feature class table. In fact, it is not uncommon to have orphan *primitives*; i.e., *primitives* that are not referenced by any *feature* at all.

```
SELECT l.name, r.alter_name, edg.coordinates
FROM "mystreet.lft" l, "mystreet.rat" r, edg
WHERE
      l.edg_id = edg.id AND
      l.name (+) = r.name
```

Figure 4.6

```
SELECT l.name, r.alter_name, edg.coordinates
FROM "mystreet.lft" AS l
      INNER JOIN edg ON ( l.edg_id = edg.id )
      LEFT OUTER JOIN "mystreet.rat" AS r
```

Figure 4.7

For the join between the feature class table and the related attribute table, the cardinality is specified in the product specifications rather than by the Standard. In our example here, we simply make the assumption that it should be a left outer join. However, it must be noted that the *feature class schema* does not have a mechanism to specify the type of joins one wants to represent.

Conversely, another SQL that expresses essentially the same operations might read as shown in Figure 4.7.

There is one very important distinction between the VPF notion of joins and the convention in SQL that we must point out. If there are multiple, e.g. two, edg rows that match one mystreet.lft row, the SQL statement above would return two result rows. In VPF, the two edg rows are combined. This is where a single street feature is composed of multiple street segments, each interrupted by intersection side streets.

If we have a spatial *merge* function in our SQL, it might read,

```
SELECT l.name, r.alter_name, merge( edg.coordinates )
FROM "mystreet.lft" AS l
      INNER JOIN edg ON ( l.edg_id = edg.id )
      LEFT OUTER JOIN "mystreet.rat" AS r
```

Figure 4.8

Now, what if a street has multiple alternative names? That remains for the database designer to answer.

Feature Join Tables

Simple *features* generally refer to those *feature classes* that are made up of one single geometric *primitive* type. For example, *features* in the class are all zero-dimension points. Or, they are all linear features or areal features.

As long as the *features* are simple and they maintain only a one-to-one, or many-to-one relationship with the underlying *primitive*, a feature join table is optional.

A one-to-one relationship here means one *feature* employs one *primitive*. A *primitive* can appear in multiple *features* in a many-to-one relationship, and we can still do without a join table.

When *features* are composed of multiple *primitives*, especially where the number is variable, a join table is generally required. Some designers refer to these kinds of features as compound features.

Feature join tables are almost always preferred in *tiled coverages*. This is because some *features* would inevitably fall on a tile boundary. Where a *feature* straddles a tile boundary, it would need to reference more than one *primitive*.

One reason that the join table is required in one-to-many *feature/primitive* joins is that there is an implicit assumption that the primitive pointer column in some *feature class* tables, the edg_id column in the last example, do not carry repeating values. In other words, this pointer column always has an item count of one.

FROM-TO INDICATOR FOR LINEAR FEATURES

The join relationship between a *linear feature* and its *primitives* has an additional qualifier. This qualifier depicts the orientation of the *feature* in terms of the underlying geometry. This allows us to construct directed graphs for specific applications. The qualifier is called the from-to indicator. It is an optional data item, however. And the from-to indicator is applicable only for *linear features*.

For example, *linear features* can be constructed into directed graphs to model a transportation network. Returning to the highway network example above, if we are to analyze the connectivity of a network, it is important that we gather up the *primitives* and arrange them in a sensible manner. The from-to indicator serves this purpose.

Although the details of *primitives* have not been discussed, let's assume that all linear *primitives* have an intrinsic direction; that is, a linear geometric *primitive* has a beginning and an end. And the intrinsic direction of travel for a linear *primitive* is from the beginning towards the terminal at the end.

Where the direction is of significance for the *feature*, a positive one (+1) in the from-to indicator means the feature travels in the same direction as the intrinsic direction of the *primitive*. Conversely, a negative one (−1) means the feature travels in the opposite direction to that of the *primitive*.

Finally, it should be noted that the *from-to* indicator is not recorded in the *feature class schema table*. That is, this qualifier is not explicitly described in the schema. It is assumed that if a from_to column is found in either the feature table or the join table for a *line feature class*, the column is automatically interpreted as the direction indicator, and the qualifier is automatically applied to the corresponding join relationship.

CHAPTER 5

Graphic Primitives

The best way to think about *primitives* in VPF is to treat them as data types that describe certain properties of *features*. Some *feature* properties are described by numbers, others by text strings. There are properties that can only be described by *primitives*. *Primitives* are just part of our modelling vocabulary.

VPF *primitives* describe the geometric properties of *features* captured in the database.

At another more elementary level, *primitives* are the data elements in a VPF database that provide the graphical representation of the database content.

While only two *feature classes* are defined and the database developer is supposed to build all his own *feature classes*, all *primitives*, including their interrelations, are fully defined in the VPF Standard. Since the implementation of *primitives* is much more controlled, there is considerably less design flexibility available to *primitives* in a VPF database than that which is available to *feature classes*. There is no equivalent to the feature class schema table for *primitives*.

Primitive tables are, by and large, predefined by the Standard. The bulk of the columns in these tables is there to maintain special kinds of integrity constraints. This mechanism to implement integrity constraints on graphical data in a VPF database is called winged-edge topology.

The end-user of a VPF database should not be exposed to *primitives* at all. The guiding philosophy in designing a VPF database should be, "Put all the interesting stuff with the features." In fact, but for a few data quality related items, *primitives* are not allowed to have any attribute. This is one of the few areas in the Standard where product-specific information is explicitly disallowed.

Besides those used to define the primitive, the only columns allowed in a primitive table are:

- optional reverse-pointers, foreign keys, to features
- meta-data columns for data quality:
      source
      positional accuracy
      up-to-dateness
      security
      releasability

Table 5.1   Primitive Types

Primitive types	Table name
Edge	edg
Connected node	cnd
Entity node	end
Face	fac
Text	txt

PRIMITIVE TYPES

There are five *primitive* types in VPF. The types are: (1) edge, (2) connected node, (3) entity node, (4) face, and (5) text. The corresponding table names for these primitive types are: (1) edg, (2) cnd, (3) end, (4) fac, and (5) txt.

Each *primitive* type is maintained in one individual VPF table. Primitive tables are placed in coverage directories. If a *coverage* is tiled, primitive tables are found in tile subdirectories. Only one table of a particular *primitive* type is allowed at a time. Hence, it is not possible to have two edg tables, e.g., in the same untiled coverage. There can be only one single independent edge, edg, table in a coverage directory, or in a tile subdirectory, if the *coverage* is tiled.

Primitives from different *coverages* are totally independent of each other. The Standard does not support any means of referencing *primitives* from multiple *coverages,* either among the *primitives* themselves or by *features.*

All primitive tables and their columns are completely defined in the Standard. The bulk of these columns and their usage reflect the winged-edge topology implementation in VPF.

Figure 5.1 shows VPF primitive tables and the relationships among them. Not every VPF database implements all the elements presented in this diagram. The diagram shows a generic *primitive* type with five subtypes. These correspond to the discussions above.

In general, edge, connected node, entity node, and face primitive types belong in the category of *geometric primitives*. The text primitive type is the only *cartographic primitive*. All these types are derived from the generic *primitive* entity. The only distinction between these two categories is that one participates in the topological structure and one does not. The integrity constraints derived from the notion of wing-edge topology apply only to the *geometric primitives*. The text *primitive* type is not governed by topology.

There is also a strange entity shown as the *ring* type. Ring tables are join tables of a sort; they allow chains of *edges* to form boundaries of *faces.*

PRIMITIVES AT LEVEL 0 IMPLEMENTATION

Figure 5.2 shows the table schema for *primitives* in a level 0 VPF dataset. Level 0 topology is the simplest of the four levels. In a Level 0 VPF dataset, only three

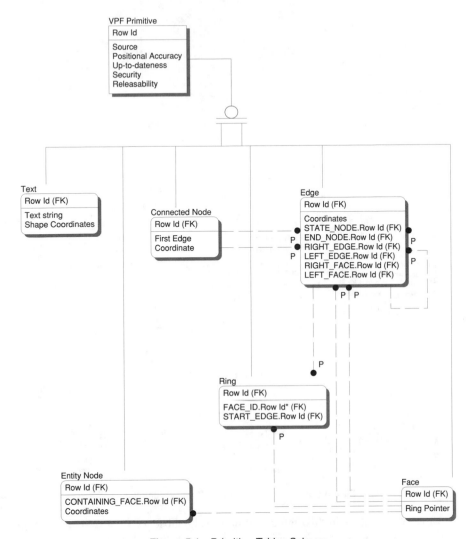

Figure 5.1 Primitive Tables Schema.

primitive types are required. These are: (a) *text primitive*, (b) *entity node primitive*, and (c) *edge primitive*. In the conceptual data model of DIGEST, *face primitives* are also allowed in level 0; however, its implementation has not been fully explained in DIGEST/VRF.

Figure 5.2 is a subset of the earlier figure. In addition to the primitive tables, a number of feature class tables are also included for discussion. As the topology level of a dataset increases, additional columns appear in primitive tables, and other tables are also added. As introduced earlier, all primitive tables are a subtype of a standard base primitive table that utilizes its row number as the primary key. Primitive tables are allowed only limited columns to support data quality implementations; these columns are shown as non-key attributes.

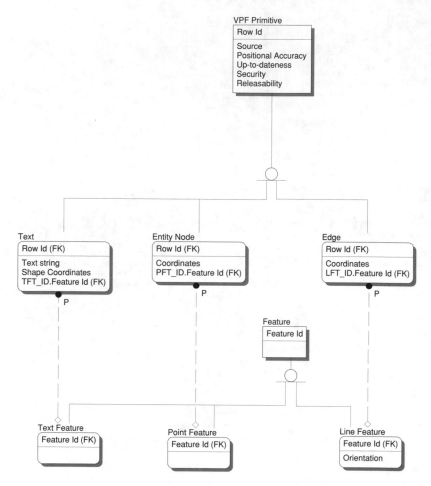

Figure 5.2   Primitives at Level 0 implementation.

Most of the tables in this figure will expand in subsequent versions illustrating higher topological levels. Text primitive tables are the simplest — they remain unchanged at all levels.

At level 0, an entity node table contains the required Row Id and coordinates columns. Where it is appropriate, it can also include a pointer column storing the identifier of the *feature* that employs the *primitive* in its definition. This back pointer is applicable only in very simple datasets; join tables, such as those described in Chapter 4, are more commonly used in a VPF database. Although only one back pointer column is shown in each of the primitive tables, multiple columns are allowed. A single *primitive* can be utilized by more than one *feature* of a particular *feature class*, and by different *features* of multiple *feature classes*.

The edge primitive table is equally straightforward at a level 0 implementation. It also contains the required Row Id and coordinates columns. However, instead of the single coordinate-pair in the entity node table, a coordinate string is needed in

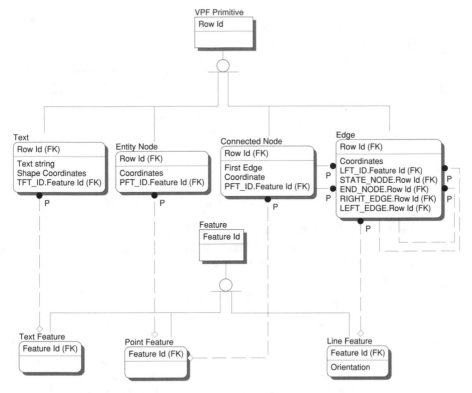

Figure 5.3   Primitives at Level 1 and 2 implementation.

the edge table. Every *edge* has at least two coordinate pairs or triplets defined in the coordinate string column.

Edges have implicit orientations. An *edge* is assumed to travel from the first coordinate pair, vertex, in the coordinate string towards the last. The orientation column in a feature class table allows a *feature* to record its own orientation vis-à-vis that of the *edge primitive*.

A relationship shown in Figure 5.2 worth reiterating is that between *features* and *primitives*. Not all *primitives* in a VPF dataset participate in feature definition. However, every *feature* must have at least one *primitive* to define its geometric property; this applies to *text features* as well.

PRIMITIVES AT LEVELS 1 AND 2 IMPLEMENTATION

In Figure 5.3, a new connected node table and some additional columns are introduced to support database implementation at levels 1 and 2. The other two tables described earlier, namely the text and entity node tables, remain the same. To make these figures more easily readable, the optional data quality columns in the primitive tables are removed.

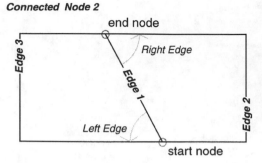

Connected Node 2

All three edges share the connected nodes 1 and 2. Node 2 is further than the end node of edge 1. Travelling counter clockwise around node 2, the right edge of edge 1 is thus edge 2.

Connected Node 1

Figure 5.4 Right/left edges in winged edge topology.

At topology level 1 and above, the beginning and ending vertices of an *edge* are made into *connected nodes*, and are explicitly stored in a connected node, cnd, table alongside the edg table. By explicitly maintaining references to these end nodes of *edges*, it helps to enforce connectivity for adjacent *edges*. If two *edges* share a common *connected node*, this relationship is clearly recognized in the database. This is different from the approach taken in the level 0 dataset. Without this explicitly maintained relationship, users of a database can only assume that two *edges* are connected if some vertices have identical coordinate values or are in close proximity of each other. Obviously, there are cases where spurious connectivity is assumed simply because two edges are close to each other. Similarly connected relationships can be lost due to coordinate values being further apart than the presumed threshold allows.

The right edge and left edge columns are also designed to maintain connectivity or adjacency relationships among *edges*. These edge columns are derived from a data structure called winged-edge topology. If one follows the direction of an *edge* from its start node towards its end node, the right edge value points to the next edge in a chain of *edges*; similarly, the left edge points to the previous *edge* in the chain. Figure 5.4 shows a simplified version of the winged-edge topology. If all *edges* incident at a common *connected node* are arranged by their incident angles, the next edge means the neighboring *edge* on the counterclockwise side.

All three *edges* shown in Figure 5.4 share the two *connected nodes. Edge 1* begins at *connected node 1* and travels towards *connected node 2*. Using the definition established above, the right edge of *edge 1* is the first *edge* encountered as one circles counterclockwise around the end node of *edge 1*. The right edge of *edge 1* is thus *edge 2*.

The first edge column in the connected node table contains a foreign key of sorts to the edge table. Since one or more *edges* can be connected to any single *connected node*, an *edge* is generally selected randomly. This column is mostly inherited from a legacy data structure that requires a pointer to navigate between an edge table and its corresponding connected node table. In a relational environment, the first edge column is of little use.

Level 1   **Level 2**

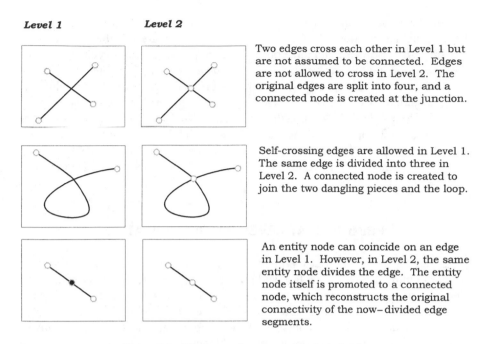

Two edges cross each other in Level 1 but are not assumed to be connected. Edges are not allowed to cross in Level 2. The original edges are split into four, and a connected node is created at the junction.

Self-crossing edges are allowed in Level 1. The same edge is divided into three in Level 2. A connected node is created to join the two dangling pieces and the loop.

An entity node can coincide on an edge in Level 1. However, in Level 2, the same entity node divides the edge. The entity node itself is promoted to a connected node, which reconstructs the original connectivity of the now–divided edge segments.

Figure 5.5   Planar requirement at Levels 1 and 2.

Levels 1 and 2 are implemented using the same primitive table schema. However, there is a planar requirement imposed on coordinate strings in edge tables by level 2 implementations. Coordinate strings in level 1 edge tables must have at least two coordinate pairs — the beginning and ending nodes. There is no other constraint on the geometric definition of an *edge*.

Primitives are required to be planar in level 2 implementation. This planar requirement means no two *primitives* can occupy the same location with a dataset; this generally means within a VPF *coverage*, or a tile, if the coverage is tiled. The planar requirement further restricts that a coordinate string cannot define an *edge* that is self crossing. Self-crossing *edges* are split at the crossing. An *entity node* also splits an *edge* if it is located directly on the *edge*. *Connected nodes* are automatically generated at ends of *edges*.

Incidentally, *text primitives* are not included in the planar requirement.

Figure 5.5 shows three situations where the planar requirement of Level 2 implementation affects *edge primitives*. In both Levels 1 and 2, *connected nodes* are defined at ends of *edges*. However, two *edges* can cross each other in a Level 1 implementation without a *connected node* at the intersection point. *Edges* are not assumed to be connected unless they share a common *connected node*. This is not allowed by the planar requirement in Level 2 implementations. *Edges* are split where they cross. The original two *edges* in Figure 5.5 are split into four and a *connected node* is created at the point where the two original *edges* cross. The newly created *connected node* hence designates the junction of four *edges*.

Similar to *edges* that intersect each other, self-crossing *edges* are allowed in Level 1 implementations. In the example shown in Figure 5.5, the original self-crossing *edge* is divided into three once planar requirement is enforced. Not unlike the case involving two intersecting *edges*, a *connected node* is created at the point where the self-crossing *edge* crosses itself. The original *edge* is therefore subdivided.

In addition to disallowing intersecting *edges*, Level 2 implementations also prohibit *entity nodes* from locating directly on an *edge*. Observing the planar requirement in Level 2 implementations, an *entity node* divides the *edge* that falls under it. The *entity node* is itself promoted to a *connected node* in the new configuration. The *connected node*, in turn, reconstructs the original connectivity of the now divided edge segments.

PRIMITIVES AT LEVEL 3 IMPLEMENTATION

Figure 5.6 shows the last topology level and its schema. Two more primitive tables are introduced here: the face and ring tables. *Edges* are connected into *rings*. *Rings*, in turn, become the boundaries of *faces*. In addition to these new tables, a number of columns are also inserted into the edge and entity node tables.

Entity nodes now have a containing face column that points to *face primitives*. Similarly, every *edge* also carries the new right face and left face pointers. All three columns contain foreign keys to *face primitives*. Null values are not allowed.

GEOMETRIC PRIMITIVES

Edge

Referring to Figure 5.1 again, the *edge primitive* table has eight columns that are of particular interest in this discussion. As noted before, these columns are used to implement the integrity constraints. Depending on the degree of topological integrity a database designer wishes to enforce, some or all of the columns might appear in a specific edg table.

As in all VPF tables, the *edge primitive* table also has the unique identifier, id, column. This simply contains a sequence number for each row.

Minimally, the *edge* table contains the unique identifier and the coordinate string *coordinate* column. The *coordinate* column gives the geometric shape of the edge. It should be noted here that the coordinate string includes the end points of the *edge* and every vertex in between.

The *start node* and *end node* columns contain foreign keys to the *connected node table*. These nodes identify the end points of an *edge*. They also establish an orientation for the *edge*; the *edge* is considered to have a direction of travel beginning at the *start node* towards the *end node*.

If we follow the direction of an *edge* as it is defined by the *start* and *end nodes*, the *right face* is simply the area to the right of the *edge*, the *left face* to the left. In

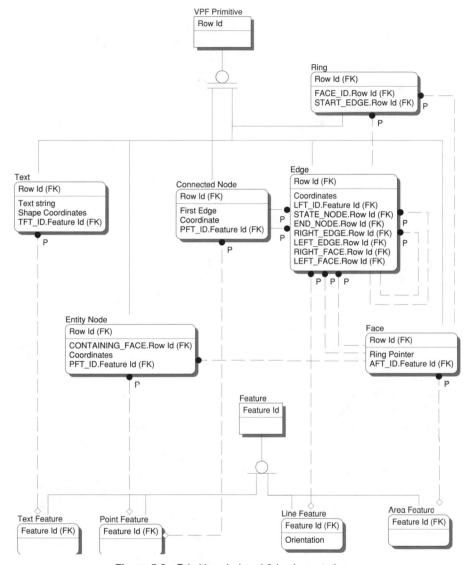

Figure 5.6   Primitives in Level 3 implementation.

Figure 5.7, for the *edge* that begins at the start node at the bottom of the diagram, the *right face* is Face B and the *left face* is Face A.

It is perfectly normal for an *edge* to have the same *face* on both its right and left sides. For example, in the dangling edge shown in Figure 5.7, Face A is identified as both the *right* and *left faces* for an edge.

If the edge is a self-closed loop, the *start node* and the *end node* columns reference the same connected node record. The orientation of the *edge* is then determined by the sequence established by the vertices.

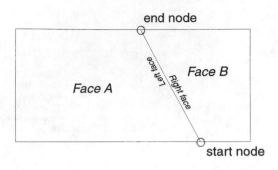

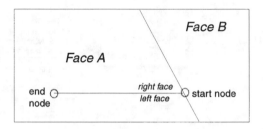

Figure 5.7 Right/left face foreign keys.

If the face topology is present, every *edge* always has one *right* and one *left* *face*. This applies even to *edges* that are found along the borders of the dataset. The *face* that is outside of the set of valid *faces* defined for the dataset is a special face called the universe face. With the universe face concept, the *face-edge* relationship is maintained throughout.

Assuming the two faces, Faces A and B, constitute our entire database, Edge 1, therefore, has relationships described earlier. On the other hand, Edge 2 is on the boundary of the database; there are no data beyond. Hence, while Edge 2 has a valid data face, Face B, on its left side, the edge has the universe face on the right.

One should also note that both edges, Edge 1 and 2, have the same *start* and *end* nodes. Obviously, these are two entirely distinct edges. This arrangement is in contrast with some other implementations of network topology, particularly those popular in transportation modelling, which use the start and end node identifiers to distinguish edges. An *edge* is uniquely identified by its id column.

The *right* and *left edge columns* are foreign keys to the edge table itself. If we place the start node of an edge and all its incident edges flat on a plane, the *right edge* is the first edge one would encounter by searching counterclockwise around the *end node*. The *left edge* is the edge, counterclockwise, around the *start node*. These two *right* and *left edge columns* implement the winged-edge topology in a VPF database.

A dangling edge might refer to itself as its own *left* or *right edge*, depending on whether the dangling end is at the start node or the end node, since it is the only edge incident at that node.

Figure 5.4 shows the winged-edge topology and the *right* and *left edges* of a very simple configuration. The *right edge* of Edge 1 is Edge 2; the *left* is Edge 3.

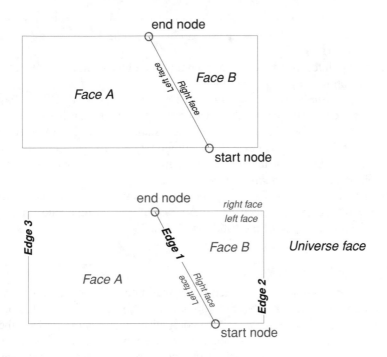

Figure 5.8 Universe face.

Connected Node

Connected nodes are found at the terminus of *edges*. These include all edge junctions and dangling ends. They can only occur in conjunction with *edges*. A separate type of *entity node* is for free-standing nodes.

Most *connected node tables* have four columns, even though only two are of any use. As is convention in VPF, there is the unique identifier, id, column. A *connected node* record contains a single coordinate pair or triplet, depending on whether the database is built using two- or three-dimensional coordinates.

The first edge column in the *connected node* table is a foreign key to the *edge* table. It stores an arbitrarily chosen *edge* from the set of *edges* that incident at the *node*. This column really serves no particular function. It was inherited from an earlier prototype of the Standard and simply remains.

The connected node table also has the infamous "optional, null" column as discussed in Chapter 3. The *containing face* column in the connected node table is not used. A *connected node* cannot be contained in a *face* within the same *coverage*.

Face

A *face* is a two-dimensional polygonal primitive. *Faces* are bounded by *edges*. Currently a *face* record does not include the coordinate string that delineates its boundary. Instead, the *face table* only maintains a pointer to the collection of *edges* that form the boundary. There is, in fact, an extra *primitive* between a *face* and the *edges*. This is the *ring*.

The *ring* pointer in the *face table* points to the outer boundary that defines the *face*. It is a foreign key to the *ring* table.

When *face primitives* are implemented for a *coverage* in a VPF database, the Standard enforces the notion of exhaustive covering. The notion is that every location within the entire geographic coverage of the database would be occupied by one *face*. Furthermore, VPF has devised a methodology to accommodate the "outside world" that is unknown or undefined in our model. This was demonstrated in the discussion on the *edge primitive* above.

The outside world is occupied by the Universe face. The first record in a *face* table is always the Universe face. The outer *ring* of the Universe face is undefined. The inner *rings* bound all of the valid data faces.

Ring

Rings are collections of connected *edges* that define the boundaries of *faces*. A *face* can have multiple *rings*, but *faces* in a VPF database must be continuous, that is, a *face* cannot have multiple disjointed components. This causes perhaps the most common confusion between *faces* and their corresponding *area features*. An *areal feature*, since it can contain multiple *primitives*, can have discontinuous *faces*.

The top diagram in Figure 5.9 shows two simple circular *faces*; each is bounded by a circular *ring*. These circles are the *outer rings* of these *faces*. This is an example of two valid *face primitives*. However, we cannot make a single *face primitive* from these two circular *rings* because the two are discontinuous.

A common, even traditional, method is to insert an artificial bridge connecting the two circles, thus making them continuous. In the old days, the "bridge" was necessary to model a chain of islands — archipelagos, for example. Because VPF *features* can be composed of multiple *primitive faces*, the spurious "bridge" is no longer necessary. Instead, if the two *faces* do indeed represent two islands in an archipelago, the *archipelago area feature*, such as the Hawaiian islands, can refer- ence these multiple *faces*. This extra burden is removed by the separation of the problem domain and *geometric primitives* implemented under strict topological constraints.

Nevertheless, a continuous face may have multiple rings. A doughnut-shaped face is bounded by two circles. The doughnut face is indeed continuous. More complicated *faces* may have any arbitrary number of *inner rings*. In fact, the *face* shown in Figure 5.9 is a continuous *face* with three *rings*. The *face* is defined by one *outer ring* and two *inner rings*. Each of the two *inner rings* defines a separate *face* — the space in the holes.

The *ring table* is essentially a join table that implements this one-to-many relationship between a *face* and one or more *rings*. Each row in the *ring table* has its (1) unique row identifier, (2) a *face* identifier, and (3) a *start edge* pointer.

Rows in the *ring* table have a predefined order. All rows for a *face* must be grouped together in the table. Every *face* will always have an *outer ring*. Depending

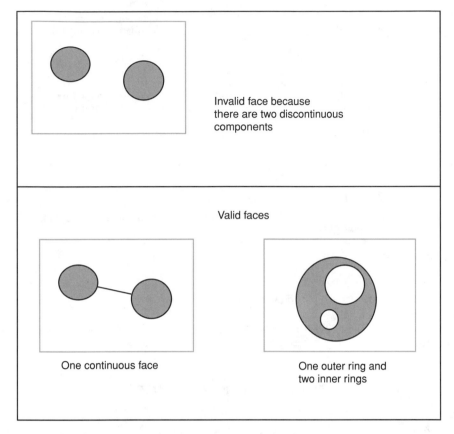

Figure 5.9 Valid face definition.

on the geometry of the *face*, there can be any number of *inner rings*. The *outer ring* comes first; the *inner rings* follow in no particular order.

The *start edge* pointer references an arbitrarily chosen edge that is part of the ring. Once one of the edges is identified, it is relatively straightforward to gather the remaining edge segments based on their *right face* and *left face* attributes.

Entity Node

Entity nodes are free-standing nodes. They are the opposite of connected nodes which are allowed only at ends of edges. However, entity node tables have the identical schema as that of the connected node tables. The only difference is that the containing face column is now valid, whereas the first edge is designated as an optional null column.

Depending on the topological constraints in effect, entity nodes may not coincide with an edge. Where face primitives are available in a coverage, entity nodes also maintain a reference to the face in which they occur. If face primitives do indeed appear

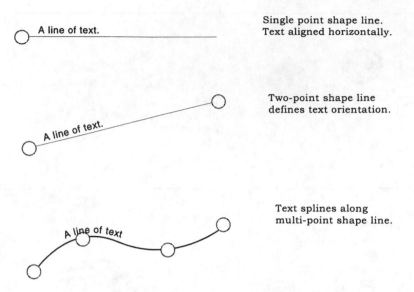

Figure 5.10   Shape lines.

in a coverage, it automatically means the highest level of topological constraints. At this highest level, entity nodes are always contained in faces.

CARTOGRAPHIC PRIMITIVES

Text

Last, we turn to the *text primitive*. The *text primitive* table has two columns, besides the other inherited from the base primitive table. The actual "string" is stored in a character string column. The intended location or placement of the coordinate string is represented in the shape line column.

The shape line column contains a coordinate string. Depending upon the number of coordinates in the string, the shape line has different interpretations. Figure 5.10 shows the use of shape lines in *text primitives*.

If the coordinate string contains one *coordinate pair*, it is generally interpreted as the lower left hand anchor for the text placement. The text is then horizontally aligned with the display window. If the string has two or more coordinates, the alignment of the text is determined by the shape line. It is possible, and is common practice, that texts are aligned with curves in addition to sloping lines.

CHAPTER 6

Topology in Geospatial Data

Topology is an extremely broad and diverse field. The segment of topology we dealt with in geographic information systems, particularly in databases developed for the VPF data Standard, is much simpler. We'll briefly review the rationale for including topology in a geographic database and examine their VPF implementation in this chapter. The topological integrity constraints in VPF are available for *primitives* only. Relationships among *features* remain entirely at the hands of product designers. All items required for its implemenation have already been dealt with in the *primitive* chapter (Chapter 5).

There are two commonly held, sometimes opposing, views regarding the use of topology in geographic information systems and their supporting databases. One predominant view held by geographic information scientists considers topological information to be redundant in a spatial database. The view holds that, generally, topological information can be derived or computed from the geometry maintained in the database.

The other view is that topological information cannot readily be reconstructed from the geometric data in the geographic database and that topological information becomes a semantic constraint on the maintenance of graphical data. Sometimes this is referred to as the finite precision geometry problem.

Supporters of the second view argue that because of the digital implementation of geometric operations in many geographic information systems, these operations suffer from the limitations of the finite precision floating point numbers, i.e., rounding errors. To ensure that data are not corrupted or lost during graphical manipulations and translations, the added topological information serves as a safeguard. Much of this stemmed from James Corbett's paper, *Topological Principals in Cartography*, a technical report written for the Bureau of the Census in 1979. Similar arguments were also developed by Victor Milenkovic (Verifiable Implementations of Geometric Algorithms Using Finite Precision Arithmetic, in *Geometric Reasoning*, MIT Press, 1989).

The second view is, by and large, being adopted in the VPF Standard. Although many might argue that the distinction is trivial, the implications from this view are important for a database developer. This is because the built-in topological information

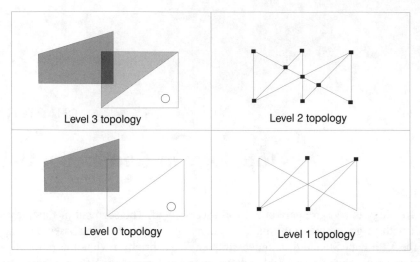

Figure 6.1   Topology levels.

in a VPF database is there only to ensure data integrity; they are not intended to directly model real-world topological relations between real-world objects. To assume, implicitly, that VPF supports applications that depend on topological reasoning is erroneous.

Remember that the topology properties specified in the VPF Standard are applicable to *primitives* only. *Features*, the real domain-specific objects, do not really have built-in topology in a VPF database. For spatial analyses that are dependent on topological in addition to metric relationships, the basic elements mandated by the VPF Standard are not likely to be adequate. If there are aspects of topology that are needed to support analyses, the database designer must include them in the database design. If an application domain requires topology, it is the responsibility of the database designer to construct those relationships at the *feature* level.

TOPOLOGY LEVELS

Figure 6.1 is a schematic depiction of the topology levels in VPF.

Level 0: Boundary Representation

At Level 0 topology, only *entity node* and *edge primitives* are available. *Connected nodes* and *rings* are constructs only applicable to higher topology levels. Level 0 topology essentially means there is no integrity constraint applied to the graphical data in a VPF dataset. Everything is determined by the geometric positions of the *primitives*.

The system does not know anything more about these *primitives* other than their *coordinates*. If one asks if two *edges* in a database meet, the result can only be

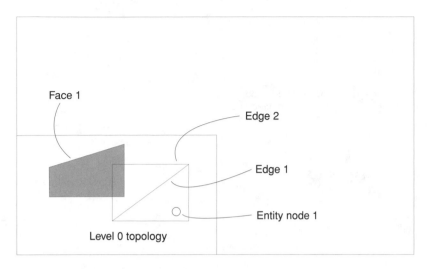

Figure 6.2 Level 0 primitives.

determined by checking the coordinates in the *edges* and seeing if we can find a coincident pair.

Figure 6.2 shows primitives in a Level 0 implementation. *Edge* 2 traces a rectangle; however, it does not define a *face*. Generally, Level 0 *primitives* simply trace outlines. The diagram is slightly misleading because VPF currently does not support *face primitives* in Level 0 topology. However, it would be straightforward to implement a Level 0 *face* using either the nonzero winding rule or the even-odd rule on an *edge*. The generic model in DIGEST does include a Level 0 *face primitive;* however, its implementation is not entirely clear.

Level 1: Linear Network

Level 1 topology adds the constraint that allows the developer to explicitly state that two particular *edges* meet. This is an explicitly maintained relationship. When we are asked the same question again whether two *edges* meet, we can respond with a true/false answer without checking the *coordinates* in the *edges*. The precision tolerance of floating point numbers in the coordinates no longer affect the query.

The position where an *edge* begins or ends is recognized as a *connected node*. If we desire to record the fact that two *edges* do meet, they both must reference the same *connected node*. Otherwise, all *primitives* can still be placed without constraint; nothing is particularly changed from Level 0.

Figure 6.3 shows the same diagram as in Figure 6.2, but with the additional Level 1 topology constraints imposed on it. All *edges* are now defined by *connected nodes* at ends. *Edges* 2 and 3, being loops, start and end at the same vertex and thus both have only one *connected node*. *Edge* 1 does have two *connected nodes*. Do note that all three *edges* cross each other at various locations. Level 1 topology allows such crossings to occur and makes no assumption about the connectivity of *edges* at those

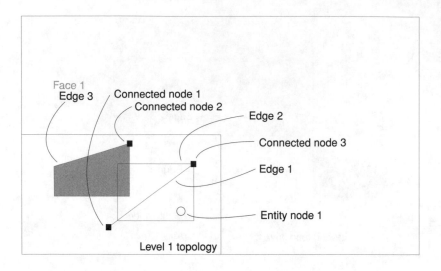

Figure 6.3 Level 1 primitives.

crossing points unless explicitly indicated by a *connected node*. Hence, *Edges* 1 and 2 are connected at *connected node* 3, at the northeastern corner.

Level 2: Planar Network

Level 2 topology enforces yet another additional constraint to the dataset. It requires that no two graphic *primitives*, of whatever type, can occupy the same position in a dataset.

In other words, if we project every *primitive* in a dataset onto the coordinate plane, every position can only be occupied by at most one *primitive*. Whereas in the previous levels, *an entity node*, as truly a free-standing node, can be placed at any position within a coordinate space, an *entity node* stored with Level 2 topological constraints can no longer be coincident with another *entity* or *connected node*, and it can no longer lie on an *edge*.

Wherever two *edges* meet, a *connected node* must be created and both *edges* must explicitly reference that *connected node* as one of its termini. An *edge* stored in a Level 2 topology cannot be self crossing, although an *edge* may start and end at the same *node*, thus forming a loop. Oftentimes, a single *edge* is therefore broken in multiple smaller pieces due to this restriction.

Figure 6.4 shows the same drawing under Level 2 implementation. Due to the planar restrictions, three new *connected nodes* are introduced along with six new *edges*.

Level 3: Planar Graph

The space bounded by the *edges* has no particular significance in a Level 2 topology. These spaces are now explicitly maintained as *faces* when the data are stored in a Level 3 topology.

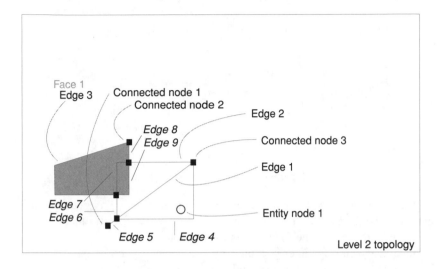

Figure 6.4   Level 2 primitives.

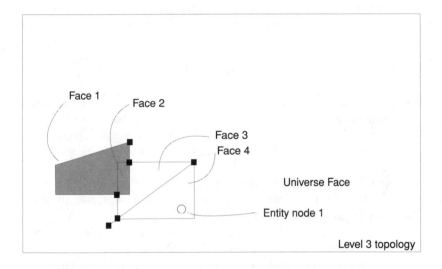

Figure 6.5   Level 3 primitives.

We can think of the complete extent of the dataset as being occupied by a universe face at the beginning. As data are inserted into the dataset and *faces* are partitioned by the bounding *edges*, parts of that universe face are being owned by these user-defined *faces* — i.e., parts of the universe face are simply being subtracted from it by the newly created user-defined *faces*. The universe face is still out there, extending to infinity in all directions. For the portion of that universe face where we have known data, we now have our own known and maintained *face primitives*.

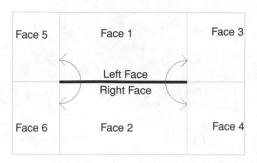

Figure 6.6   Winged-edge topology.

Figure 6.5 shows the five nonoverlapping, complete cover *faces* occupying every location of the dataset. All the *edges* and *connected nodes* are still there — they are not drawn to make sure the figure remains readable.

Since everything is either occupied by the universe face or by a user-defined *face*, *entity nodes* now have an explicitly maintained relationship with the *face* that contains it. *Entity nodes* still cannot be coincident with *connected nodes* or *edges*.

Throughout our discussion we have not mentioned the *text primitive*. Topological constraints are not applied to *text primitives*. *Text primitives* can be located anywhere. They can lie on or under any other *primitives*, including another *text primitive*.

IMPLEMENTATION

VPF employs a data structure called winged-edge topology to implement topological constraints in the database. Winged-edge data structure was first introduced by Baumgart in 1975 (Baumgart, B.G., 1975, A polyhedron representation for computer vision, *Proceedings of the American Federation of Information Processing Societies,* 1975).

Baumgart's winged edges did look like edges with wings. Figure 6.6 shows the winged-edge data structure. There are 6 faces drawn in the diagram; there are also 13 edges. For the purpose of this discussion, the *edge* of interest is the one at the middle, drawn in thicker black. This *edge* goes from left to right, thus it is bounded on the right by *Face* 2. The wings refer to pointers to neighboring *edges* which, together with our *edge*, define the boundaries of these *faces*. In Baumgart's design, there are four such pointers for each *edge*.

The winged edge in VPF is a slightly modified version of Baumgart's. Figure 6.7 shows the VPF winged-edge data structure. Instead of four pointers, there are only two pointers in the VPF implementation.

The complete VPF data structure for winged-edge topology thus involves: (1) start node, (2) end node, (3) right face, (4) left face, (5) right edge, and (6) left edge.

We have already been introduced to all the items that are needed to implement the topological integrity constraints in a VPF database. Columns that we described in the *primitive* chapter (Chapter 5) are mostly developed to enforce these constraints.

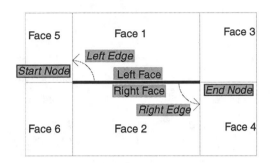

Figure 6.7 Winged-edge topology in VPF.

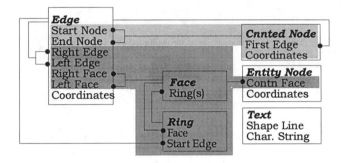

Figure 6.8 VPF topology.

As noted in that chapter, dependent upon the level of constraints the database developer desires, some or all of the columns might be included.

Figure 6.8 shows the tables and columns constructed for VPF topology. The shadings in the diagram are used to indicate the level of topology — the darker the shading, the higher the topology level. These columns and tables are incremental. That is, columns and tables are added to the minimal set as one increases the level of data integrity one wishes to enforce on a dataset. At topology Level 0, only three tables and four columns are required (Figure 6.9).

Levels 1 and 2 topology essentially have the same implementation items. Figure 6.10 shows the tables and columns for Level 2 topology. The number of tables increases to four, and the number of columns to ten. Furthermore, we start to encounter relationships being maintained among the *primitives*.

The only difference between Level 1 and Level 2 topology, as stated earlier, concerns whether *primitives* can overlap. The most obvious distinction appears at the intersections of edges. In Level 1 topology, two edges are allowed to cross each other without automatically creating a connected node at the intersection point.

At Level 2 topology, every intersection point automatically generates a *connected node*. *Entity nodes* are no longer allowed to be located on an *edge*, or be coincident with a *connected node* or another *entity node*. In fact, at Level 2 topology, no two *primitives* are allowed to overlap.

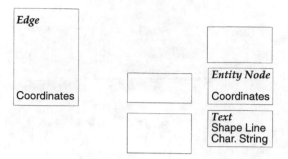

Figure 6.9 VPF topology Level 1.

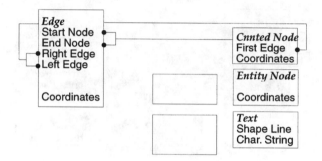

Figure 6.10 VPF topology Levels 1 and 2.

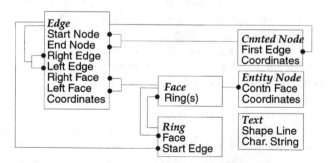

Figure 6.11 VPF topology Level 3.

The entire complement of tables and columns introduced in Chapter 5 are employed only in Level 3 topology datasets. The difference between Levels 2 and 3 topology is the treatment of the space delineated by the edges.

In Level 2 topology, the spaces between *edges* are not explicitly encoded. These spaces are explicitly defined as faces in Level 3. Figure 6.11 shows the Level 3 topology in VPF.

LIMITATIONS

Finally, it must be reemphasized here that the implementation of topology in the VPF Standard is only sufficient to assist in enforcing some aspects of data integrity for the graphical elements in the database. The built-in topology is not adequate for spatial analysis of even the simplest kinds.

Wherever the application domain requires topological information — properties such as connectivity, adjacency, and so on — the database designer must explicitly implement these properties for the applicable *feature classes*.

CHAPTER 7

Thematic Index

Indices in a VPF database are generally divided into four categories: (1) thematic index, (2) spatial index, (3) join index, and (4) variable-length table index. Most indices are optional depending upon the application or the database browser to take advantage of thematic and spatial indices available in a VPF database. However, few databases are produced without accompanying thematic and spatial indices.

Thematic indices are available for all data types with the exception of coordinate-string data types. Spatial indices are designed specifically for these data type columns. All indices in VPF are, by and large, defined on single columns. Join indices are incorporated into VPF to support operations on relational tables. A last category of indices contains variable-length table indices. The variable-length table index is defined for entire records rather than individual columns. In addition, a variable-length table index is the only type of index that is mandatory.

VARIABLE-LENGTH TABLE INDEX

A variable-length table index is designed to enhance the processing of a table with records that can be of variable lengths. For tables that have only constant-length records, it is a simple calculation to position at any record in the table during I/O operations. However, it can be much more expensive to process a table with variable-length records. This index allows a browser to access arbitrary records in the table without costly processing of individual records.

A table is considered to be a variable-length table if it satisfies either of the following conditions: first, if any of the *count* fields in the column definitions of the table schema allows a variable number of items in the column, and second, if the table includes a *triplet id* type. If either of these conditions appears in a table schema, that table is a variable-length table. Variable-length table indices are automatically required for these tables.

As stated in an earlier discussion (Chapter 3), the file that stores a variable-length table index and the index itself are named after the source table for which the index

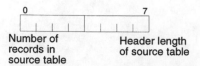

Figure 7.1 Variable-length table index.

is built. The file holding the index is distinguished from the source by means of the file extension. More precisely, the index file is distinguished from the source mainly by the last character of the three-character file extension. With one exception, the last character in the file extension of a variable-length table index is the character "x". The rest of the filename is copied from the filename storing the source table.

Unlike the other thematic indices that are explicitly declared in a feature class schema, variable-length table indices are not explicitly defined in the database schema. This is because these indices are always required for the conditions introduced above, and are named according to the specified naming convention.

Variable-Length Table Index Header

The variable-length table index is divided into two parts. First, the index header — there is only one record in the header. The first field in the header record stores the number of records in the corresponding source table, for which the variable-length table index is built. This indicates the number of entries that appear in the second part of the index table.

The second field in the index header record stores the header length of the source table. This information could easily be obtained from the source table directly. Figure 7.1 shows the layout of the variable-length table index header.

Variable-Length Table Index Data Entry

Following the header record, the second part of the index table contains a number of entries, as specified by the value stated in the header. The variable-length table index is itself made up of fixed-length records. Each record has two long integer fields. Figure 7.2 depicts the data entries in an index.

Each entry in a variable-length table index stores the starting position and the length of a record in the source table. Every record in the source table is represented with one entry. Therefore, entries in the index table are maintained in a one-to-one relationship with source table records. Entries and records are additionally maintained in the identical order; hence, the first entry refers to the first record, the second entry refers to the second record, and so on.

The starting position of a record in a source table is stored in the first field of an index entry. This position refers to a location in the file that stores the source table. The position is measured as a byte offset; this offset is measured from the very beginning of a file. The first byte of a file has a byte offset of zero. Even though the header length of the source table is stored in the index, source table record positions are still measured from the very beginning of that file.

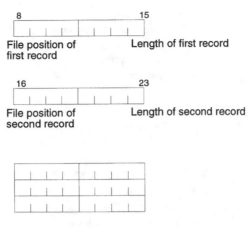

Figure 7.2   Variable-length table index.

THEMATIC INDEX

Thematic indices in VPF databases are designed principally to enhance attribute queries against a VPF database. As database designers of VPF products realize more and more of the relational database concept in their product design, thematic indices take on the increasingly important role of supporting relational joins.

A thematic index can only be defined on a single column. VPF does not support the construction of multicolumn indices. Conversely, a column can have only one index defined on it. In addition, thematic indices can only be defined on nonspatial, nonrepeating value columns. Thematic indices are not applicable to spatial columns — columns that contain coordinate strings.

Although columns that contain character strings do store repeating-value fields, these columns are allowed to have thematic indices under certain restrictions.

Thematic indices are explicitly named. Recalling the earlier discussion, column definitions of a table schema can optionally include thematic index references. Unlike variable-length table indices, filenames of thematic indices are not required to mimic the corresponding name of the source table. Thematic indices are stored in separate, individual files. These files are placed in the same subdirectory as their corresponding source table.

There are two types of thematic indices available in VPF databases. These are the inverted list and the bit-array indices. The inverted list index type is the more common of the two. An inverted list provides a simple mechanism to access records in a VPF table via ordered searches instead of a complete table scan. A bit-array index is more likely to be used for character-string columns. It can quickly filter a large number of possible candidates by means of a pattern-matching strategy.

Thematic Index Header

A thematic index can be divided into three sections: (1) the header section, (2) a directory section, and (3) a data entry section. The header section gives the overall

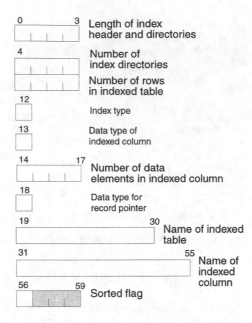

Figure 7.3 Thematic index header.

information about the index. Copies of indexed values are maintained in the directory section; it is a lookup table of sorts. The data entry section contains references back to the original source table for which the index is built. These references are, by and large, record numbers.

Figure 7.3 shows the structure of the thematic index header. The header section of a thematic index is always 60 bytes in length. It provides basic information about the index.

The first header field contains a byte count of the length of the header and the directory sections of the index. This is a *long integer* value.

The second and third fields give information about the source table. The second field contains a *long integer* value that represents the number of directories in the index. The notion of index directories is slightly different for the two index types. The third field is a *long integer* value that stores the number of records in the source table for which the index is built.

Next comes the index type field, which indicates whether the index is an inverted-list or a bit-array. Inverted-list indices are given a type value of "I"; bit-array indices are indicated by a "B".

The fifth field records the data type of the column that the thematic index applies. This matches the data type as declared for the column in its column definition.

The next *long integer* field stores the number of data elements in the indexed column. It might seem curious that this field exists at all in the header because only nonrepeating value columns can have a thematic index. The field is reserved for future extension of the Standard. This field is currently always given a value of one (1), except for fixed-length character-string columns. For fixed-length character-string

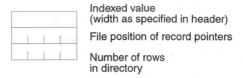

Figure 7.4   Thematic index directory entry.

columns, the number of data elements represents the length of the character string reserved or declared for the column.

Record pointers or references are stored in the third, data entry section. Either a short integer number or a long one can be specified for record pointers. If the largest value for the record pointers can be represented by a short integer number, using a short instead of a long record pointer obviously reduces the size of an index file by half.

The next two entries in the header give some more background information related to the index. These two fields contain the name of the indexed table followed by the name of the indexed column.

The next byte includes the character "S" to indicate that the directories are sorted. Directories can only be sorted in ascending order. If the byte contains any other value, index directories are assumed to be unsorted. The last three (3) bytes in the header are reserved and should remain empty.

Inverted-List Thematic Index

Directories in an inverted-list index store unique, or distinct, values found in the column for which the index is applied. One index directory is created for each distinct value found in the column. The order in which these directories appear in the index is determined by whether directories are sorted or not. Directories are sorted by the distinct values. Thematic indices in a VPF database only support sorting in ascending order. It is difficult to imagine a justification why the directories should remain unsorted. Nevertheless, index directories are not required to be sorted.

Each directory in the index is made up of three parts. Figure 7.4 shows this index directory structure. The first part of the directory stores the indexed values. The second part is a pointer to a location in the thematic index table where record references are maintained. These locations are actually sequences of integers, either *short* or *long integers*, that store record numbers of records which have the corresponding value in the indexed column. The number of records that share the same distinct value is stored in the third part of the index directory.

The scheme just presented is called the indirect addressing scheme. If more than one record in the source table shares the same value, indirect addressing is the only applicable method. When a directory contains only one record entry, there are two different methods to store the record number. The record number can be stored using the indirect method as described here, or it can be stored using the direct addressing method.

This method is called indirect addressing because one gets to the index directory and, using the pointer in the second part of the directory, jumps to another location

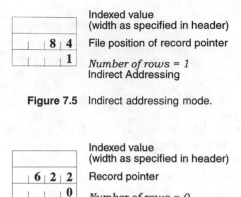

Figure 7.5   Indirect addressing mode.

Figure 7.6   Direct addressing mode.

to pick up the record number. Figure 7.5 shows an example of indirect addressing in a thematic index. With the indirect addressing method, the record number of that single record in the directory is still stored in a separate index entry in the data entry section of the index table. A value of one (1) is entered in the third part of the directory. This seems like unnecessary computation. With indirect addressing, the accessing program first processes the directory record, then by following the pointer in the second part of the directory, retrieves the single record number from the data entry section.

That same record number of the single record can be stored, indeed, in the second part of the directory entry, thus saving the extra processing step. This is the direct addressing method. In the direct addressing method, the record number is placed in the second part of the directory, and the count in the third part of the directory record is set to zero (0) instead of one, as it would be normally. A direct address example is shown in Figure 7.6.

Both address schemes are possible in a single index table. There is no requirement that direct addressing must be implemented, although it is possible to achieve some efficiency gain by using the method.

Example of Inverted-List Thematic Index

Figure 7.7 shows an example of an inverted-list index. The source table is an arbitrary table named a1.aft. This table has five records. As shown in the index header, there are five records with four distinct values.

Figure 7.8 shows an implementation of an unsorted inverted-list index. There are four directory entries. The entry for "3" is shown at the first directory entry because the value 3 is the first value encountered while iterating through the a1.aft table. Because the value 3 appears in two records, the count in the directory entry shows 2. Following the pointer in the directory entry, the record numbers 1 and 3 are found. The other three directory entries are relatively straightforward. Each of these is a single-entry directory.

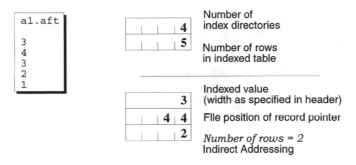

Figure 7.7   Inverted list example.

File position of record pointer

Number of rows = 2
Indirect Addressing

Unsorted directories (indirect addressing)

3	4	4	2
4	1	0	4	1
2	2	8	4	1
1	4	6	4	1

		1		3
		2		
		4		
		5		

Figure 7.8   Inverted list index implementation.

File position of record pointer

Number of rows = 2
Indirect Addressing

Sorted directories (indirect addressing)

1	4	6	4	1
2	2	8	4	1
3	4	4	2
4	1	0	4	1

		5		
		4		
		1		3
		2		

Figure 7.9   Sorted index implementation.

Figure 7.9 shows a sorted inverted-list index on the same source table. Everything appears pretty much the same except that directory entries are now arranged in an ascending order according to the values on which the index is applied.

Figure 7.10 shows a further variation of the sorted inverted-list index on the same table. In this last implementation, the direct addressing scheme is used where applicable. Thus, all directories which have a single record entry are now assigned a count of zero (0), and the record numbers are stored directly in the second part of the directory.

File position of record pointer

Number of rows = 2
Indirect Addressing

Sorted directories (direct addressing)

1	5	0
2	4	0
3	4 4	2
4	2	0

1	3

Figure 7.10   Direct address implementation.

Bit-Array Thematic Index

Bit-array thematic indices are mostly applicable to support text pattern searches. Bit-array indices share the same index header structure as the one used in inverted-list indices. A bit-array index has an index type code of "B". This appears in the index type field in the index header. The rest of the index table, however, is quite different from the other index type.

Bit-array thematic indices still have a directory section. Although directory entries are also made up of three parts, the pointer — the second part — is replaced with a bit-array, thus giving the name of the index type.

The bit arrays are of constant length. Each record from the source table is represented by one bit. Hence, the length of the arrays reflects the number of records in the source table. Bit arrays are aligned to byte boundary. Therefore, whether the source table contains ten or fifteen records, the array length is still two bytes long.

Bits in bit arrays are in one-to-one correspondence with records in the source table. The first bit represents the first record, the second bit the second record, and so on. Bit-array indices do not implement the notion of direct and indirect addressing schemes. To determine the appropriate record or record set, positions of the appropriate bits in the bit array are used. Each bit is either on or off, depending on whether the pattern specified for the directory is observed in the indexed column in the corresponding record. Navigating a table by means of a bit-array index is simply the application of bit-wise operations on these arrays.

The third part in directory entries as described earlier in the inverted-list indices is the same for bit-array indices. The values given to the number of records in index directories are identical for all directories in a bit-array index. A better use of this count field is to record the number of bits. More efficient processing strategies can take advantage of the information to plan the bit-wise operation necessary to obtain the desired records. Since the bit-array thematic index is hidden far beneath the other more exotic elements of a VPF database and, indeed, bit-array indices are not used very often, this change most likely would not be implemented in the foreseeable future.

t1.tft
ASSUA
USA
ABBA
USSB

A	**1**	**1**	**1**	**0**				
B	**0**	**0**	**1**	**1**				
S	**1**	**1**	**0**	**1**				
U	**1**	**1**	**0**	**1**				

Figure 7.11 Bit-array index example.

Bit-Array Index Example

The following example illustrates one application of a bit-array index in support of a text pattern search on a column in the now familiar t1.tft table. Figure 7.11 shows four records of the table. A variable-length text column is identified as the search attribute. The text string for this column in the first record has a length of 5. The lengths range from 3 to 5.

Luckily, a bit-array index has been constructed for this column. The directory section of this index is shown in Figure 7.12. Taking each letter as a unique pattern, there are altogether four distinct patterns representing the four different letters that appeared in the strings in this column.

In searching for a particular pattern using a bit-array index, it becomes a simple process of finding the intersection of the various components that make up the pattern.

A series of simple bit-map operations on the index directories is all that is required to locate the records which contain the string "USA" in the t1.tft table using the t1.tft index.

Reviewing the tables in Figure 7.12, t1.tft has four records. Each of the records contains a character string field. In this highly simplified example, values in this field are various combinations of four letters of different lengths; the lengths range from a three-character string to one with five characters. The four letters are: '"A", "B", "S", and "U".

The four letters become index directory entries in the bit-array index. Each of the directories has a four-bit-long bitmap, or bit array, representing the four records in t1.tft. Since the letter "A" appears in the first, second and third records in the table, their corresponding bits in the "A" index directory are turned on — that is, assigned a value of 1. There is no "A" in the fourth record; hence, its bit is turned off.

In searching for a particular pattern from a bit-array index, it becomes a simple process of finding the intersection of the various elements in the pattern. For example, to locate the country name record that is "USA", the search begins with selecting an arbitrary letter in the search pattern and retrieving the appropriate bit array from the index.

In this example, the letter "U" is found in records one, two and four. The initial bit-array thus returned would have those three bits turned on.

The letter "U" is found in items one, two, and four.

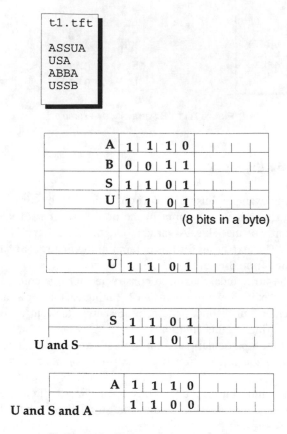

Figure 7.12   Bit-array index processing.

The second bit-array returned is the one from the "S" directory. The letter "S" also appears in records one, two, and four.

The intersection of these two bit arrays thus returns the same candidate list.

The last operation is to intersect the result obtained from the first two arrays with the bit array from the "'A" directory.

The intersection of these two returns only two on-bits indicating that the first and second records in the table can potentially satisfy the match pattern.

This methodology only reduces the solution space rather than return the exact matches. As shown in this example, the bit-array search does not consider the order in which the components appear in the match pattern, nor does it handle multiple occurrences of a component within a pattern. A follow-up check for exact matches is generally necessary to obtain the desired result.

FEATURE JOIN INDEX

The relational database notion is fundamental to the implementation of VPF. In fact, in DIGEST/VRF, it explicitly recognizes the relational database concept in the format title. When designing a relational database, normalization and relational joins inevitably crop up. Conventional single-column thematic indices have taken on an increasingly important role in supporting these relational operations. Another index type, the feature join index, is constructed specifically to perform the prejoining of feature class tables and their component geometry tables.

The VPF Standard specifies join indices for feature/primitive joins only. Currently, the feature join index cannot be used for other table joins in a database.

Due to the inherent nature of the VPF database, i.e., the separation of geometry and attributes, feature join tables are highly recommended. VPF databases, as explained above, are relational in nature. VPF databases are also essentially static in nature. These databases are published for querying and browsing. Few applications require high interactive editing and modification of the database. Maintaining the set of prejoin results in index tables thus becomes highly desirable with minimal penalty.

Join indices* are developed to enhance the processing of complex queries in relational databases. The performance improvement is particularly significant in partitioned, distributed databases. The notion of tiling in VPF databases parallels that of fragmentation in distribution databases. For a more complete treatment of the underlying theory and performance benchmarks, Valduriez's paper, though dated, is an excellent reference. Any of the data warehousing discussions would also provide a sufficient background examination of the join index.

A feature join index in VPF essentially performs a prejoin among the feature class tables and the primitive tables in a coverage. The resultant index stores all feature/primitive pairs in the coverage. These pairs include both direct feature-to-primitive relationships, such as line features that use edge primitives, and indirect references between complex features and primitives.

Direct feature-primitive relationships are straightforward enough. Indirect ones are more complicated and it is here that the join index becomes essential. In an indirect reference, a complex feature may utilize both node and edge primitives in its definition. Using the structure supported in the VPF Standard, a waterworks complex feature includes both aquaduct and pumping stations. Aquaduct features arc defined using edge primitives, whereas the pumping stations are node primitives.

For an edge feature join index in this coverage, entries must be created for the aquaduct feature-to-edge pairs, and, for the waterworks, feature-to-edge pairs which reference the edge primitives indirectly using the aquaduct features as go-betweens.

Join indices are defined at the coverage level. This handles the partitioning of the database in tiled coverages. However, feature join indices are applicable for both tiled and untiled coverages.

One feature join index can be defined for each of the five primitive types in VPF. An edge feature join index, edg.fit, is for edge primitives in a coverage and all the corresponding line and complex features that reference those primitives. Similarly there are the connected node feature join index, cnd.fit, and end.fit for entity nodes. The face feature join index, fac.fit, is for face primitives. The text feature join index, txt.fit, is also available but does not appear as commonly as the others.

* Valduriez, P. 1987. Join indices. *ACM Transactions on Database Systems*, Vol. 12, no. 2 (June), 218–246.

The use of the the feature join index follows the one-for-all and all-for-one principle. If an index is deemed desirable for one primitive type in a coverage, all other primitive types in that coverage must also have join indices.

If a join index is built for one primitive type, then all feature classes that utilize that particular type of primitive must be included in the join index.

Join Index Table Schema

Feature join indices are stored by using regular VPF tables. Also the join index uses a separate tile id column rather than the triplet id to access the primitive. There can be five such tables; each maintains an index for a primitive type. The VPF table that contains a feature join index must contain the following columns:

- Feature class ID
- Feature ID
- Primitive ID
- Tile ID (if the coverage is tiled)

Feature Join Index Example

Figure 7.13 presents a feature join index example. There are four feature classes in this example coverage. The feature classes include one point class and two line classes. In addition, there is also a complex feature class. These feature classes are built using edge and entity node geometric primitives. For the purpose of this example, the coverage does not enforce any topological integrity constraint.

The feature classes are labelled "c0" for the complex feature classes. The line feature classes are called "l1" and "l2"; while the point feature class is "p1". These are shown in the feature class attribute (fca) table. None of the feature class tables shows any attribute column. All joins are applied on the id column. This is sufficient to illustrate the join index.

The simplest of these joins are found in the l1 line feature class and the p1 point feature class. Reading the feature class schema (fcs) table in Figure 7.13, features of the l1 line feature classes are simple single edge features. To construct any of the l1 line feature, it is simply a matter of joining the id column in the l1.lft table and the id column of the edg table. The first l1 line feature thus has edge-1 as its geometry. This can be obtained easily enough from the feature class and primitive tables.

The join index table also records these relationships. Edge primitives 1, 2, and 3 are shown in the first three rows in the edg.fit feature index table. The middle column of these three rows shows a 2 for these rows. A feature class id of 2, reading from the fca table, indicates the l1 line feature class. The third column shows the feature ids.

Thus interpreting these rows, the first row says edge primitive 1 is used by feature 1 of feature class 2, the l1 line feature class. The second row shows edge primitive 2 used by line feature 2, and the third row has edge primitive 3 and line feature 3.

The same description can be applied to the first three rows in the entity node feature index table. For simple, one-to-one feature-to-primitive joins, the join index table does not really demonstrate a tremendous efficiency gain. The advantages here

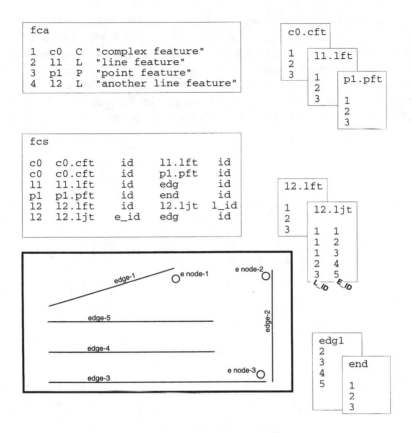

Figure 7.13 Join index example.

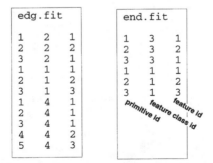

Figure 7.14 Join indices implementation.

are that, given any primitive, it is straightforward to identify all the features that reference the particular primitive. Without the feature join index, it is necessary to append various redundant backpointers to the primitive table.

The most apparent advantage of the feature index table is where joins are relatively more complicated, such as in the case of complex feature classes. The complex feature class, c0, is composed of the l1 line feature class and the p1 point feature class. In order to construct a feature in the c0 complex feature class, the component line and point features must be built first. After this has been accomplished, the component features are joined together to make up the complex feature. This processing sequence is particularly expensive for these operations to be performed in every draw operation or query. By performing prejoins and storing the results in the feature index tables, a simple one-step lookup is all that is required.

CHAPTER 8

Spatial Index

Spatial indices in VPF include both minimum bounding rectangle (MBR) tables and spatial index tables. These indices are utilized in spatial searches and graphical displays. During searches, a spatial index allows a more intelligent path to navigate the dataset. Indices also enhance graphical displays by strategically reducing the set of features and primitives that must be retrieved, processed, and displayed. Spatial indices are applied on VPF primitives only. The VPF Standard does not prescribe a spatial index for its features.

All spatial indices in a VPF database are implicitly defined; that is, they are not specified in a schema or header table. All spatial indices have predefined reserved names. Since spatial indices are solely associated with primitives, index tables are located where their corresponding primitive tables are found.

The spatial indices shown in Figure 8.1 are defined in the VPF Standard.

MINIMUM BOUNDING RECTANGLE (MBR)

The minimum bounding rectangle (MBR) is the simplest indexing scheme for geometric primitives in a VPF database. It also provides the basis for the slightly more complicated semiquad tree spatial index.

As its name suggests, a minimum bounding rectangle is the smallest possible rectangle that completely contains a particular edge or face primitive. Minimum bounding rectangles in VPF are always aligned along the horizontal and vertical axes in which the coordinates are defined in a database. For databases that maintain spatial objects in geographic coordinates, this implies that all bounding rectangles are aligned with the longitudes and the latitudes.

Minimum bounding rectangle tables are mandatory for edge and face primitives. A minimum bounding rectangle record is required for each record in an edge or face primitive table. Node and text tables do not have bounding rectangles. An edge minimum bounding rectangle, ebr, is needed for an edge primitive table, whereas a face minimum bounding rectangle table, fbr, is required for the face primitive table. These always show up in pairs.

Primitive types	MBR Table name	Index Table name
Edge	ebr	esi
Connected node		csi
Entity node		nsi
Face	fbr	fsi
Text		tsi

Figure 8.1   Spatial indices.

```
Edge Bounding Rectangles;
;
id=I,1,P,Row ID,-,-,-,:
xmin=F,1,N,Minimum x-coordinate,:
ymin=F,1,N,Minimum y-coordinate,:
xmax=F,1,N,Maximum x-coordinate,:
ymax=F,1,N,Maximum y-coordinate,:;
```

Figure 8.2   MBR Schema.

MBR Table Format

The VPF Standard defines one single MBR table format. This single definition is applicable to both edge and face minimum bounding rectangle tables, the only two types currently supported.

A minimum bounding rectangle table is a regular VPF table, which is slightly different from most other indices in a VPF database. The table contains as many records as the number of primitives in its associated primitive table.

Figure 8.2 shows the table schema for the bounding rectangle table. Each record in a MBR table contains five (5) mandatory columns. The column definitions in the figure illustrate an edge bounding rectangle table. The five fields are the ubiquitous id columns. This is accompanied by four *float* columns which store the bounding coordinates.

Coordinates that define a bounding rectangle in the MBR tables are always stored in *long float* or *float*, depending on the data type that has been chosen for the primitive table. Minimum bounding rectangles do not support the z-coordinate, even though primitives in the index table might include the third dimension.

MBR for Universe Face

Each entry in the minimum bounding rectangle table references one geometric primitive. Records in both tables follow the identical order. Hence, records in the edg and the ebr tables parallel each other. The same applies to records in the fac and the fbr tables.

The relationship between a minimum bounding rectangle record and its corresponding primitive record is one-to-one always. This, however, introduces a rather tricky situation with the face minimum bounding rectangle table.

The minimum bounding rectangle of a face primitive is defined using the outer ring of the face. As presented in earlier discussions, every face primitive table references a "universe" face in its first entry. The universe face is an artifact derived from the topological constraints enforced on VPF databases rather than some real geometry that has a physical definition. The face is assumed to extend to infinity; consequently, its bounding rectangle is not very meaningful as an aid in processing VPF databases.

Whether the bounding rectangle for the universe face is meaningful or not, there remains the issue of how to encode the rectangle in the face minimum bounding rectangle record. One suggestion is to exclude the universe face MBR from the minimum bounding rectangle table. Yet another is to define the bounding rectangle of the universe face using its interior rings instead.

In order to maintain consistency in the definition and an established data production convention, the first entry in a face minimum bounding rectangle table, fbr, continues to refer to the universe face. The minimum and maximum coordinates in the record are specified as *null* by the Standard. By and large, VPF database readers should treat the bounding rectangles of universe faces simply as undefined.

SEMIQUAD TREE SPATIAL INDEX

The semiquad tree indexing process consists of three fundamental steps; they are normalization, allocation, and division.

The first step in spatial indexing is normalization, which transforms the normally floating point numeric values of coordinate strings into integers. In addition, the coordinate system in an index coverage is normalized to an integer range between 0 and 255, inclusive. Another way to visualize this normalization process is to consider the entire coverage being divided into 65,536 cells, with 256 cells across and 256 up and down.

The second step, allocation, and the third step, subdivision, are performed iteratively. At the beginning of the allocation process, all 65,536 cells belong to bin 1. A primitive is allocated into a bin which completely contains the minimum bounding rectangle of that primitive in normalized coordinate space. It is obvious that everything will automatically be allocated to bin 1, but bins have limited capacities.

When a bin is allocated beyond a predetermined threshold number of primitives, it is subdivided. The threshold is called the bucket size.

The third step, subdivision, divides a bin into two halves. The root of the index tree, bin 1, covers the entire extent of the indexed coverage. Thus it ranges in the x axis from 0 to 255; similarly, the y axis ranges from 0 to 255.

The first subdivision is along the x axis. Subsequently, the subdivision alternates between the y axis and the x axis. The second level therefore has one branch that

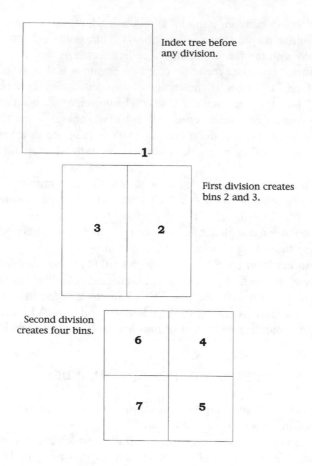

Figure 8.3   Semiquad tree index.

ranges along the x axis from 0 to 127 and another that ranges from 128 to 255. Both branches have a y axis from 0 to 255.

Figure 8.3 shows the division algorithm. All second level bins have the same dimension. The second level bins have half the width of their parent. However, they have the same height as that of the parent bin.

Subdivision and allocation are repeated until all bins are within their capacities; that is, the numbers of primitives allocated to bins is less than the specified bucket size for the index. After the first division, primitives are reallocated into the smaller halves as well as the parent bin if they saddle the halves. The threshold limits of the bins are again checked.

Spatial Index Bucket Size

The bucket size of a semi-tree spatial index specifies the number of primitives a spatial indexing algorithm would allow in a bin before the bin must be subdivided.

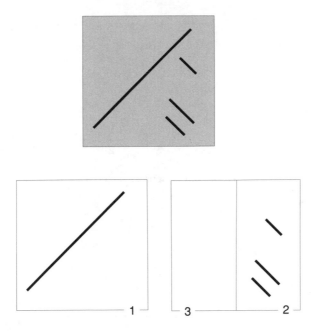

Figure 8.4 Subdivision steps 1 and 2.

The optimal number is governed principally by the applications that will make use of the spatial index. The VPF Standard does not prescribe a bucket size. Instead the value of the bucket size is generally only defined by a product specification. A bucket size of 8 has been found to be rather typical and practical in most VPF databases.

Division Algorithm and Numbering Scheme

The diagram in Figure 8.4 shows a number of snapshots through the index tree. This is a schematic representation of the division and numbering scheme. In reality, all these bins are stacked on top of each other. If, indeed, a slice is cut from the index tree, all bins in a level would have uniform size.

The first node of the tree, bin 1, covers the entire extent of a dataset. The first division is applied vertically; the original bin is separated into right and left halves. The right half is assigned bin 2 and the left bin 3. These two are not shown in the diagram.

The next division is aligned horizontally; this division is applied simultaneously to bins 2 and 3. Bin 2 is subdivided into two halves. Bin 4 is the top half. Bin 5, as shown in the diagram, represents the bottom half of bin 2 after the division.

Bin 3, the original left half of the second-level division, is similarly divided. Bin 6 is the top half; it is shown in the diagram (Figure 8.6). Bin 7 is the bottom half but it is hidden in the diagram.

The next division is again aligned horizontally. Rather than a single cut as in the level two division, this division includes two cuts. The first cut separates bins 4

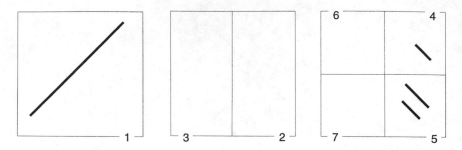

Figure 8.5 Subdivision step 3.

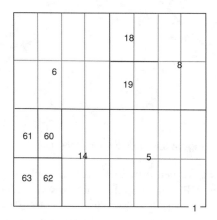

Figure 8.6 Semiquad numbering.

and 5 into 4 bins; the other cuts bins 6 and 7. Bin 8 is obtained from bin 4. Bin 8 is the right half, and bin 9 the left half. Cell 5 gives cells 10 and 11, while cell 6 has bins 12 and 13. Bin 14, which also appears in the diagram, occupies the right half of the earlier bin 7. The left neighbor of bin 14 is bin 15.

Bins 8 and 14 are shown in the diagram.

This division process repeats until one of two conditions is met. The division process stops when all primitives are allocated into bins and the bins all have sufficient capacities to handle the allocated primitives. Or, the division repeats until bins have a width and height of one unit; that is, each bin covers only one cell. This is rare because the threshold limit, the bucket size, can be adjusted.

The fundamental rule is to put into a subdivision only those primitives that are entirely within the range of that subdivision. The number of graphic primitives one can put in each of the subdivisions is determined by the bucket size.

The diagram presents index bin 1, the root, as a square, and all subsequent cells as rectangles arranged in portraits.

This shape is entirely the artifact of the normalization process. This shape does not reflect the geographic footprint of the original dataset.

```
Subdivide( level, bin num, xmin, ymin, xmax, ymax )

if (( xmin = xmax ) and ( ymin = ymax )) then Cannot Subdivide end if
if ( ++level = odd ) then
        left child := ( bin num * 2 ) + 1
        right child := ( bin num * 2 )
        Subdivide( level, left child, xmin, ymin, floor( xmin+xmax /2 ), ymax )
        Subdivide( level, right child, ceil( xmin+xmax /2 ), ymin, xmax, ymax )
else
        bottom child := ( bin num * 2 ) + 1
        top child := ( bin num * 2 )
        Subdivide ( level, bottom, xmin, ymin, xmax, floor( ymin+ymax /2 )
        Subdivide ( level, top child, xmin, ceil( ymin+ymax /2 ), xmax, ymax )
end if
```

Figure 8.7   Numbering algorithm.

```
Level 5 (Bin: 060) [ x: 032..063] [ y: 064..127]
Level 5 (Bin: 061) [ x: 000..031] [ y: 064..127]
Level 4 (Bin: 030) [ x: 000..063] [ y: 064..127]
Level 5 (Bin: 062) [ x: 032..063] [ y: 000..063]
Level 5 (Bin: 063) [ x: 000..031] [ y: 000..063]
Level 4 (Bin: 031) [ x: 000..063] [ y: 000..063]
Level 3 (Bin: 015) [ x: 000..063] [ y: 000..127]
Level 2 (Bin: 007) [ x: 000..127] [ y: 000..127]
Level 1 (Bin: 003) [ x: 000..127] [ y: 000..255]
Level 0 (Bin: 001) [ x: 000..255] [ y: 000..255]
```

Figure 8.8   Bin definition.

Figure 8.8 shows a decomposition of a branch of the index tree. It traces the decomposition that results in the four bins found at the lower left-hand corner in the diagram. These are bins 60, 61, 62, and 63.

The figure shows a partial result of the subdivision and numbering operations. The resultant bin definitions show both the normalized coordinates and the bin numbers.

As explained earlier, bin 1, the root of the index tree, is divided into two halves. The left half is bin 3. Bin 3 is, in turn, subdivided into a top and a bottom half. The top half is bin 6, as shown. The bottom one is bin 7.

Bin 7 becomes bins 14 and 15. Bin 15 is the only level 3 division being tracked in this example. Bin 15, being the left-hand side of the division, is itself divided into bins 30 and 31.

Bins 60 and 61 are obtained from bin 30. Bin 61 starts from the leftmost boundary, thus its x coordinate begins at 0. It covers up to 31 along the x axis. Bin 60 covers 32 to 63. Both of these bins have the same vertical extents; their y coordinates cover 64 to 127.

Last, Figure 8.9 shows the overall structure of the divisions. Before any division is applied, there is one bin. Since a bin remains on the index tree even after it has been subdivided, there are altogether three bins after the first division. There are, in total, 17 levels based on the scheme defined in the VPF Standard. At the last division, 131,071 bins are accumulated. However, seldom are all 17 levels utilized in an index.

	Bins at level	Total bins in index	Subdivision
Level 0	1	1	No partition
Level 1	2	3	Vertically
Level 2	4	7	Horizontally
Level 3	8	15	Vertically
Level 4	16	31	Horizontally
Level 5	32	63	Vertically
Level 6	64	127	Horizontally
Level 7	128	255	Vertically
Level 8	256	511	Horizontally
Level 9	512	1,023	Vertically
Level 10	1,024	2,047	Horizontally
Level 11	2,048	4,095	Vertically
Level 12	4,096	8,191	Horizontally
Level 13	8,192	16,383	Vertically
Level 14	16,384	32,767	Horizontally
Level 15	32,768	65,535	Vertically
Level 16	65,536	131,071	Horizontally

Figure 8.9   Operations.

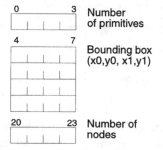

Figure 8.10   Spatial index header.

Because the number of subdivisions is finite, there is still the chance that an index might use up all the divisions. The bucket size specified for the index needs to be modified where an index is unable to allocate all primitives into the bins.

Figure 8.9 shows a summary of the subdivision and numbering operations. It also shows the number of bins generated at all levels as the semiquad tree is being built.

SPATIAL INDEX HEADER

The header starts with a *long integer* field which contains the number of primitives in the indexed table. Figure 8.10 shows the layout of the spatial index header.

The bounding box in the spatial index header is represented by using *short floating* point real numbers irrespective of the original data type used to encode

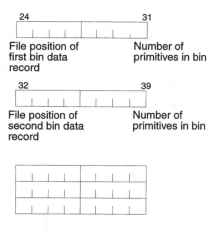

Figure 8.11   Bin array.

coordinate strings in the primitive table. This bounding box is a minimum bounding rectangle that contains all primitives in the dataset being indexed.

The bounding box coordinates are maintained in the original coordinate reference system. Only the coordinates in the bin data records are normalized. This bounding box defines the normalization parameters. The minimum x and y coordinates of this bounding box are translated into zeros. The maximum coordinates are translated to 255.

The number of nodes in a spatial index is determined by the height of the index tree; it is the number that appears in the third column in Figure 8.9. This number is not a count of the bins in which primitives are indeed present. The number of nodes value is represented by a *long integer* in the header.

Data Offset

The bin array record contains two fields (Figure 8.11). The first field stores a data offset. The offset for spatial index bin data is computed for file positions following the bin array. The first byte after the bin array has a byte offset of 0. This is different from the offset definition in thematic index tables.

The second field contains the number of primitives allocated to that particular bin. It is the number of things collected in the bucket. Also, unlike the two different addressing strategies in the thematic index, there is no direct addressing in a spatial index. If the number of primitives has a value of zero, it means no primitive is allocated to the bin.

Bin Data Records

Bin data records contain one copy of an 8-byte data structure for each of the primitives found within the bin. This structure includes a *primitive id* and the bounding box for the primitive in normalized coordinates. The bounding box coordinates appear in the first four bytes, with the *primitive id* in the next four. The

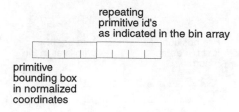

repeating
primitive id's
as indicated in the bin array

primitive
bounding box
in normalized
coordinates

Figure 8.12 Bin data records.

Bin 1

	0		1
	8		3
	0		0

Figure 8.13 First subdivision.

primitive bounding box is stored using normalized coordinates. They fit into the 4-byte field since each coordinate can be represented by a single byte. The layout is shown in Figure 8.12.

Because tiled coverages have different spatial indices for different tiles, primitive references in spatial indices do not require the *triplet id* data type and they are not prefixed with a tile qualifier.

SPATIAL INDEX EXAMPLE

Let's review a simple example of how an index can be constructed. The indexed coverage contains four edge primitives. After normalization and the initial allocation, all four primitives are placed into bin 1.

Assuming an arbitrarily set bucket size of 2, which is entirely unrealistic, bin 1 is overflowing. This bin is thus subdivided into bins 2 and 3 using the algorithm presented earlier.

There are three bins in the second-level subdivision; hence, there are three bin-array records (see Figure 8.13). Bin 1 contains one primitive. This primitive crosses the two smaller bins.

Bin 2 contains the remaining three primitives. Bin 3 still appears in the bin array even though it contains no primitive. Since it contains no primitive, the primitive count field in the record has a value of zero. As required by the Standard, the offset field also has a value of zero.

The number of primitives allocated to bin 2 is still the predetermined bucket size of 2 after the division. Another level of subdivision is therefore required.

In this new subdivision, bins 2 and 3 are divided into four smaller bins. The entire index now has seven bins — thus the seven records in the bin array.

The three primitives that were allocated to bin 3 originally are now separated into bins 4 and 5. The allocation of bin 1 remains unaltered.

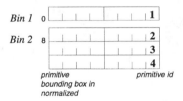

Figure 8.14 Second subdivision.

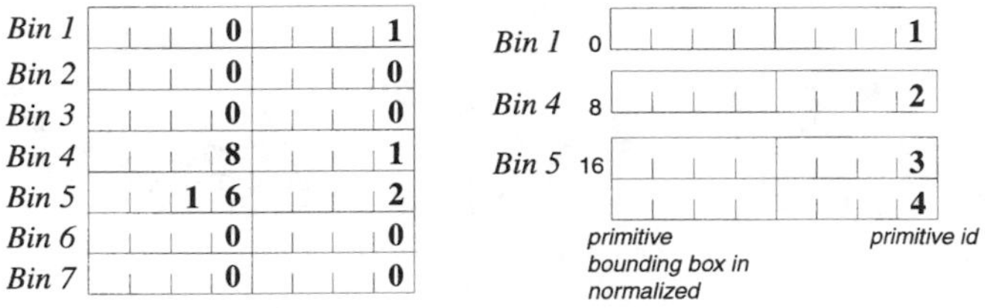

Figure 8.15 Third subdivision.

Bins 2 and 3 are now both empty. Bins 6 and 7 are empty as well. Nevertheless, all four empty bins are still represented by their bin array records. The bin array and records thus become as illustrated in (Figure 8.16).

NAVIGATION WITH SPATIAL INDEX

A common application of the spatial index is during a spatial query. A user-specified coordinate is given, and the query is to identify the feature which the user points at.

This process would first normalize the input coordinate. It is then straightforward to locate a bin for this query point. By and large, the bin at the smallest subdivision is selected. Bounding rectangles of primitives found within the bin are compared with the point.

For those successful candidates, that is, those bounding rectangles that contain the query point, their actual geometric definitions are compared with the query point. This more precise comparison is generally performed in the original, unnormalized coordinate reference system.

If no match is found for the spatial query, the search space is expanded to the level above in the index tree. This is repeated until a match is found or until the tree has been exhausted.

LIMITATIONS

The discussions in the last two chapters generally do not concern the data users or a VPF database modeller. Few would ever get to examine the internal format of a spatial index in a VPF database.

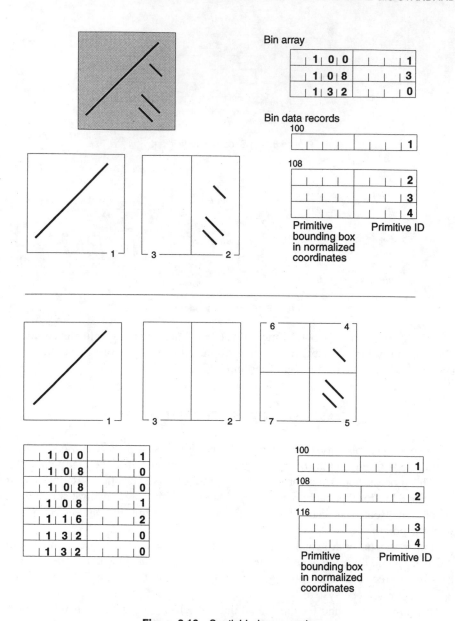

Figure 8.16 Spatial index example.

For a database designer it is generally beneficial to include as many indices as possible, given limitations in production and manufacturing costs. A semiquad tree spatial index is generally recommended for every primitive table. The minimum bounding rectangle tables are, nonetheless, required.

By and large, it is dependent upon the application or the database browser to take advantage of thematic and spatial indices available in a VPF database. Few users would actually need to know the exact format and structure of a VPF index as described in the last two chapters.

CHAPTER **9**

Tiling

Tiling is simply another method of enhancing data processing in VPF databases. Much like *spatial indices*, tiling is a mechanism to provide better performance in accessing data. Tiling is basically a divide-and-conquer strategy in handling large spatial databases.

In a relational database, a table can be partitioned both horizontally and vertically. Vertical partitions mean that columns in a table are divided and rearranged into multiple separate tables. Horizontal partitions are also rearranged but rows are subset instead. *Feature classes*, in some instances, are examples of vertical partitions. However, partitioning a database is almost always an entirely physical design consideration. *Feature classes*, with their intrinsic meaning in the spatial database, are generally results of logical rather than physical or administrative design consideration.

Tiling, on the other hand, is a horizontal partition of VPF *coverages* that does not alter any of the logical meaning of the data. Tables being partitioned in this case are *primitive tables* in the *coverage*. Tiling transforms a regular VPF *coverage* into a *tiled coverage*. Every *primitive table* in the *coverage* is transformed, i.e., partitioned. Tiling also creates multiple subdirectories in the original coverage directory; each subdirectory contains all the primitive tables that appear in the original coverage directory.

One very important point to note is that tiling in VPF does not affect *features* and *feature classes*. On the other hand, each individual *primitive table* in the original *coverage* is partitioned into multiple *tiled primitive tables* having the same structure as the original. It is not uncommon to create empty *tiled primitive tables*. These *tiled primitive tables* are placed into the appropriate tile subdirectories. *Tiled primitive tables* retain the original table and file names of the previous unpartitioned version. The only change in the database schema relates to the primitive identifiers. More precisely, it is related to the use of primitive identifiers as references, or foreign keys.

In untiled coverages, primitive identifiers uniquely identify every *primitive* in a *coverage*. There are two alternatives regarding these identifiers when a *coverage* is tiled. Identifiers can remain pretty much unchanged. That is, primitive identifiers remain unique across tiles in *tiled coverages*. This would have made life much easier than the alternative that has been chosen for VPF.

The VPF Standard requires that primitive identifiers be the same as row ids. Thus, primitive identifiers are unique only within a *tile*, i.e., a partition, in a *tiled coverage*. By choosing this alternative, two rather nasty consequences result. First, every *primitive* is renumbered. By itself, this is not too bad, although it does result in a cascaded update. Second, primitive identifiers are no longer unique in a *coverage*; they are only unique in individual *tiles*. To find a *primitive* in a *tiled coverage* thus requires the use of a tile prefix to fully qualify the identifier for the *primitive* in a *tiled coverage*. This ultimately leads to the unnecessarily complicated *triplet id* data type.

DATA STRUCTURES FOR TILING

Tiling in VPF is implemented using three data structures. The first is the *tile reference*, tileref, coverage. There is only one *feature class* in this *coverage*; it is the *tile reference area feature class*. The *tile reference area feature class*, also tileref, describes the tiling scheme for a *library*. All *coverages* in the *library* share the same tiling scheme.

The second data structure is comprised of tile subdirectories. Tile subdirectories are located beneath the coverage directory for a *tiled coverage*. *Primitive tables* are placed in these subdirectories. All associated *spatial indices* and *variable-length table indices* for these *primitive tables* are also placed in these tile subdirectories.

The third element is the cross-tile topology that enforces topological integrity on *primitives* across tile boundaries.

As topological constraints are applied to *primitives* in regular *coverages*, the same set of constraints also applies to *primitives* in *tiled coverages*. In order to accommodate this operation, the original set of topological constraints is supplemented by cross-tile topology.

TILEREF COVERAGE

The tiling scheme for a VPF library is defined by a tileref *coverage*. The tileref name is a reserved coverage name. Whenever a tileref *coverage* is encountered in a *library*, all other *coverages* in that *library* are assumed to be tiled according to the tiling scheme defined in that tileref *coverage*.

In a *tiled library*, all *coverages* except the tileref *coverage* are tiled. The tileref *coverage* itself is a standard *untiled coverage* with level 3 topology. In the tileref *coverage* there must exists a tileref *area feature class*. It is actually these *features* that define the tiling scheme.

Tileref *Area Feature Class*

Each tileref *area feature* in the *coverage* represents one *tile* in the partitioned *coverages*. Tile partitioning in VPF is performed according to geometry. That is, tile area feature boundaries define boundaries of database partitions. These boundaries define an area where *primitives* from the corresponding *tiled coverages* are allocated.

There is no explicit restriction on the number, size, and geometric configuration of these *tile area features*, which leads to rather intriguing situations.

For example, if *tile area features* overlap, a particular *primitive* can then appear in more than one *tile*. Similarly, if there are gaps in the *tile reference coverage*, some *primitives* might not find a home in the *tiled coverage*.

Pragmatically, the standards seem to imply that these tileref *area features* are to maintain one-to-one relationships with the underlying *face primitives*, except for the universe face. This solves the overlapping concern since no *primitive* can overlay another in a level 3 topology. However, the coverage issue still remains. The VPF Standard does not explicitly stipulate a coverage extent for the *tile reference coverage*. This one item is left for the database designer to resolve.

Tiling is a divide-and-conquer approach. *Tiles*, in general, should be of a size that they benefit in their performance in query and in retrievals. However, if *tiles* are too small, a certain amount of caching is lost. By and large, tiling in VPF should be utilized as another indexing mechanism. Because tile boundaries are defined by the database designer, it is possible to partition a *library* in such a manner to optimize access within a *tile*. Some products also allow the distribution of individual *tiles* rather than the complete *library*. Configuration of tile boundaries thus moves beyond purely storage and caching considerations.

Finally — returning to the original concern about the spatial coverage of the tiling scheme — it is obvious that a tiling scheme should spatially encompass all *primitives* in the *library*, otherwise, *primitives* might disappear in the tiling process.

Tileref *Area Feature* Attributes

Attached to these area features are two mandatory attributes; these are: (1) a unique id, and (2) a tile_name column. The unique identifiers are the primary keys here and are used as foreign keys in tile_id columns or in *triplet ids*.

There is no other column required in the *tile reference area feature*, tileref.aft, table. The unique tile id is used as the qualifier for primitive identifiers in the *tiled coverage*. Although it is called tile_name, the tile_name column actually stores directory references to the tile subdirectories.

Face Primitives in the Tileref *Coverage*

The geometry of tileref *area features* are, of course, defined by their corresponding *face primitives*, the same as any other *area feature* in any VPF *coverage*.

Although it is possible to define a tileref *area feature* that contains multiple *face primitives*, few designers have pursued this approach. Moreover, as mentioned above, the Standard assumes that tileref *area features* maintain a one-to-one relationship with *face primitives*. Since most tileref *area features* are, indeed, relatively simple *area features* that utilize only one *face primitive*, it is common to see a fac_id column in the tileref area feature attribute table rather than the more complicated join table implementation.

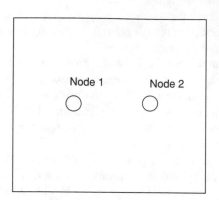
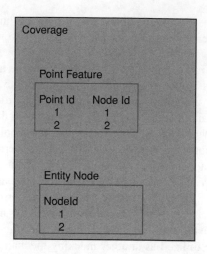

Figure 9.1   Tiling example.

However, this approach does enable tremendous flexibility in the physical design of a VPF library. Every *coverage* within a VPF library must use the same tiling scheme, but the *coverages* need not use the same *tiles*.

Hence, if a scheme containing 16 squares divides a rectangle extension of a *library*, each of the square faces can indeed represent a *tile*. Groups of twos or fours can also be made into *tiles*. In fact, it is perfectly legal to have a *tile* that covers all 16 square *faces*, thus creating an untiled *coverage* in a tiled *library*.

In another example, a tiling scheme can be constructed from counties or prefectures. There are county or prefecture tiles as an immediate result of this design. However, counties can be grouped into states and prefectures into provinces, thus resulting in a set of different tiles.

By allowing different aggregations of tiling in a library, very densely populated *coverages* in a library might use smaller tiles than the more sparsely populated *coverages* in the same *library*.

TILING EXAMPLE

The simplest example to visualize this partition is to imagine a *coverage* containing free-floating *entity nodes*, thus eliminating the complicating factor of topological constraints for the time being. A *point feature class* is defined on the *nodes*.

Features and *nodes* have a one-to-one relationship. Before tiling there is one feature table and one primitive table. Figure 9.1 shows the two points and the tables. In an *untiled coverage*, both the *feature class table* and the *primitive table* are located in the same directory.

A tiling scheme that divides this *coverage* into two halves would result in two tile subdirectories beneath the original coverage subdirectory. The original single *primitive table* is also partitioned into two tables, which are placed in their respective tile subdirectories.

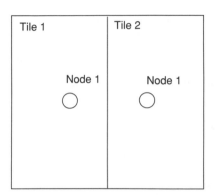

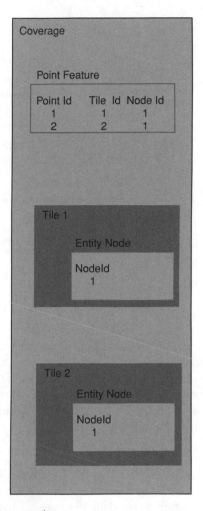

Figure 9.2 Tiling example.

Although there is no subdivision of the *feature table*, changes are still unavoidable there. Formerly, each *feature* in the *point feature table* maintained a reference to its corresponding *node primitive*. The reference needs to be updated as a result of the tiling operation. To do this a tile qualifier must be attached to the primitive reference.

There are two ways to achieve this: the original *integer* reference can be replaced by a *triplet id* reference, or a tile_id column can be added to the *feature table*. Either method accomplishes the task of constructing a qualified tile primitive reference. The latter approach, albeit a slightly kludgy one, is the more common approach among existing databases. The tile_id column name has an implicit interpretation in VPF databases. It is important to remember that this column is not explicitly declared in the *feature class schema*.

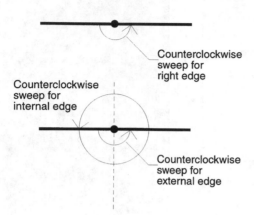

Right edge pointer expanded to include
an internal right edge and an external
right edge.

Counterclockwise
sweep for
right edge

Counterclockwise
sweep for
internal edge

Counterclockwise
sweep for
external edge

Figure 9.3   Cross-tile topology.

Figure 9.2 shows the tables after the coverage has been tiled. The single coverage directory now contains two tile subdirectories, each with its own *node* table. The *feature class table* is modified to use a tile id quantifier to locate a *primitive*.

Point features and *entity nodes* are the simplest arrangements because neither demand sophisticated topological constraints. When higher level topological constraints are present in a coverage, additional considerations are consequently required in preparing the tiled version.

CROSS-TILE TOPOLOGY

Although tile definitions are relatively straightforward, the implementation of a tiled coverage demands special attention. The topological integrity constraints introduced earlier need to be expanded to accommodate the results of *primitives* being separated into different tiles. *Entity node primitives* are relatively easy to handle.

Unlike conventional geographic databases, topological constraints that govern the integrity of VPF elements ensure integrity across tile boundaries. Tiling in VPF is a horizontal partition that is truly a lossless transformation.

Figure 9.3 shows the cross-tile topology implementation for *edges*. In this example, two *edges* meet at a *connected node*. The *right edge* is simply the first *edge* encountered on a counterclockwise sweep at the *end node*. Assume that a *tile* now divides these two *edges*, and the division just so happens exactly at the *connected node*.

The original forward edge pointers are modified to include an internal and an external component. The external is still the *edge* that was pointed to earlier albeit

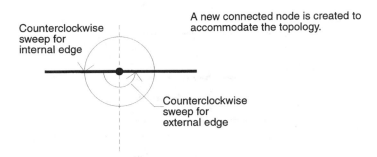

Right edge pointer expanded to include
an internal right edge and an external
right edge.

Counterclockwise
sweep for
internal edge

A new connected node is created to
accommodate the topology.

Counterclockwise
sweep for
external edge

Figure 9.4   Cross-tile topology.

located in a tile outside of the one the host edge is found. Now assume the *tiles* are divided, and the right half has been pulled away. The same sweep at the same *end node* now encounters another *edge*; in this case, it finds itself on a full circle trip. The newly identified neighboring *edge* is the internal right edge.

Figure 9.4 shows a more common situation where a single edge happens to be divided by the tile boundary. The scenario remains fairly much the same. The internal and external edge pointers at the end node are still identified in the same manner.

However, in this situation, a connected node is actually created. Recall that connected nodes are automatically created at ends of edges. Since the tile boundary breaks one single edge into two, the point where the tile boundary intersects the original edge generates a connected node. In fact, two connected nodes are generated — one to terminate the edge on the left tile and another for the edge on the right tile. The internal and external edge pointers are identified using the same procedure described. Both newly connected nodes have their own edge pointers. Incidentally, the external edge referred to by one node is the internal edge of the other.

In summation, one must do the following, to construct the internal and external edge references. Before *primitives* are actually separated into different tiles, a *connected node* must be present wherever an *edge* intersects a tile boundary. If a *connected node* does not already exist at the intersection of the *edge* and the tile boundary, a new *connected node* must be created at that location.

Regardless of whether the *connected node* is an existing one or one created by the tiling process, this *connected node* is eventually duplicated on both of the *tiles*. The external edge is the forward *edge* obtained at this *node*, before the two *tiles* are pulled apart. The internal edge is identified after the two neighboring tiles are separated and the *primitives* have been independently rearranged into separate tiles.

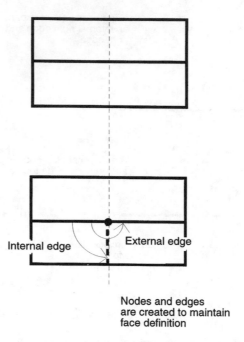

Figure 9.5   Cross-tile topology.

The strategy is followed when a *face primitive* is divided instead of an *edge primitive*. However, in addition to the *connected node* at the intersection of the *face ring* and the tile boundary, a new *edge* is also created to maintain a complete *ring*. As before, the *external edge* is the *forward edge* before the two tiles are pulled apart and the new *ring* is configured. After the tiles are pulled apart and new *edges* are inserted to repair the *ring*, the internal edge is found. Figure 9.5 illustrates this operation with two simple *faces* being divided by a tile into four *faces*.

SUMMARY

Tiling in VPF is a horizontal partition of a *database*; it is truly a lossless transformation. Generally, tiling should be considered as another method much like indexing to enhance processing performance. This lossless transformation is accomplished by the *cross-tile topology* extension to the *wing-edge topology* and the infamous *triplet id*.

The paramount rule in VPF tile processing is that *features* cannot be created or destroyed as a result of tiling. Thus, by extension *feature tables* are not partitioned by tiling. Whether a *coverage* is tiled or not, there is only one *feature table* for each *feature class*.

Primitive tables, on the other hand, are partitioned and placed in separate tile subdirectories as a result of a tiling operation. In addition, new *primitives* are often constructed during the tiling process.

When a single *primitive* is broken into multiple components due to a tile boundary, all of the resultant *primitives* should remain accessible to the *feature* that is defined on them, before tiling.

When a *primitive* is duplicated due to coinciding with a tile boundary, both replicas should also be accessible to the *feature*. To make sure duplicated *primitives* are retrieved properly, pointers are expanded to include internal and external components. Only one of these components should be in use at a time.

CHAPTER 10

Data Quality

While neither VPF nor DIGEST/VRF specifies performance criteria for data quality, the Standard provides various means to maintain quality information with a database. VPF and DIGEST/VRF do, however, provide various mechanisms for the encoding of data quality information as part of its large metadata support. DIGEST has an additional data quality information structure to store data quality information about an entire transfer package. A VRF dataset might be a subset of such a package.

DATA QUALITY TABLE

The most common, and most sharable, method is the use of *data quality tables*, dqt, and lineage description tables, lineage.doc. The *data quality table* in particular shares many of the elements specified in other geographic data and library metadata schema, including the Federal Geospatial Data Committee (FGDC), Content Standard for Digital Geospatial Metadata. VPF has, by and large, observed the fields and reports specified by FGDC, as well as the encoding and interpretations established by the various Standard committees that developed the metadata standards for geospatial data.

Figure 10.1 shows the table layout for the *data quality table* in VPF. Since the *data quality tables* can be placed at all levels in a VPF database, the first two items identify the structural level and the particular data element within that level. For example, a VPF level could be a *database*, a *library*, a *coverage,* or a *tile*. The data element name corresponds to the *database* name, a *library* name, the name of a *coverage,* or a *tile*.

Lineage information is usually reported in free-formatted text. This type of recording is best encoded using the narrative table in VPF. Conventional approach is to use the narrative table associated with a dqt for lineage information.

FGDC recommends the lineage information to include a description of the source material from which the digital data were derived or captured. It also recommends that the method of data capture be included.

Data Quality Table

Row Id
VPF Element level
VPF Element name
Feature completeness
Attribute completeness
Logical consistency
Edition number
Creation date
Revision date
Name of product
Date product spec published
Date of earliest source
Date of latest source
Std Dev of quantitative attribute
Reliability of qualitative attribute
Name of collection spec
Name of included source file
Abs horizontal accuracy
Unit of measure for abs horiz acc
Abs vertical accuracy
Unit of measure for abs vert acc
Point to point horiz accuracy
Unit of measure for relative horiz acc
Point to point vert accuracy
Unit of measure for relative vert acc
Comments

Figure 10.1   Data quality table.

Where the source material was transformed during the production process, all transformations involved should be reported in the lineage information. Among these many possible transformations are the mathematical transformations of coordinates — cartographic projections. FGDC also suggests that the coordinate transformation algorithm, locations of any registration points and sample computations be included. Recalling the geographic reference table — the registration points and diagnostic points tables for VPF libraries — those tables describe the final transformation of the data. Numerous other transformations might have applied to the data before the final carto-graphic transformation described in the geographic reference table for a library.

Since data quality and lineage information can change from coverage to cover-age, or even from tile to tile, data quality tables are often simultaneously imple-mented at multiple database levels. General information is provided at the higher database and library levels. Progressively more detailed information is provided at the lower coverage and tile levels.

Although the Standard specifies the required items within the data quality table, VPF does not specify where these tables are mandatory. Implementation of dqts remains at the discretion of a product specification. However, at a minimum, data quality information should be reported at the *database* and library levels in a VPF *database* to provide an overview of the data source, information capture, production, and maintenance processes.

FEATURES AND PRIMITIVE DATA QUALITY INFORMATION

A second approach is to store data quality information with individual *features* and *primitives*. This approach involves simply the addition of data quality columns to *feature* and *primitive tables*. The Standard is very restrictive when dealing with its *primitive tables*. However, a limited set of data quality information columns is allowed to be appended to *primitive tables*. Figure 5.1 shows the allowed items. They are:

1. source
2. positional accuracy
3. up-to-dateness
4. security
5. releasability

A product specification is still required to define the interpretation of these columns. The security and releasability fields are arguably not data quality information. There is no specific guideline concerning the use of data quality columns in *feature tables*.

DATA QUALITY COVERAGE

These two more conventional approaches in the maintenance of data quality information in geospatial databases are oftentimes inadequate to fully communicate data quality information in a VPF database. How would a producer indicate that a particular quadrant in the database coverage is especially lacking in data? A lengthy textual description in the standard data quality table would not be very precise.

The Standard includes a special purpose *coverage* for the storage of data quality information. The *data quality*, dq, coverage is a regular level 3 coverage but is especially recognized by the Standard because it maintains data quality information related to other datasets and coverages.

This third approach of communicating data quality information about a VPF database is possibly the most intriguing one. The notion of a data quality coverage or overlay is that it is possible to communicate spatially distributed quality information about data in a *library*.

The *data quality coverage*, dq, is just a regular *coverage*. There is no difference between this *coverage* and any other *coverage* that stores data in a *database*. *Quality coverages* are found in *libraries*. Because the data quality coverage is just another coverage in a library, it is conceivable that various spatial analyses can be performed using the quality coverage as quantifier regarding the validity of the analytical model or as indicator of parameter sensitivity. Of course the *coverage* also serves as an excellent visual reference of quality information when other data are displayed. Although it seems a very exciting concept, it has not been fully mandated by the Standard and few other Standard committees outside of DGIWG.

The general recommendation from the Standard is to define a data quality area feature class in the coverage. Each of these area features delineates regions of similar quality characteristics or source material. Attributes attached to the data quality area features reflect mainly the DIGEST metadata coding scheme. These columns include:

- Attribute accuracy. Attribute accuracy describes the accuracy or reliability of attribute data within the limits described by feature completeness. Attribute accuracy is to be expressed in standard deviation of attribute value, according to the requirement stated in DIGEST. If attribute accuracy information is not available in the above form, a description of known attribute accuracy characteristics may be substituted.
- Attribute completeness. Attribute completeness refers to the percentage of feature attribute fields not populated by null or default values. It is 100% complete if all the relevant attributes of a feature are captured given a capture criterion.
- Date status. Date status refers to the date at which the data was introduced or modified in the database. This date of entry is used as a proof of modification for a single data element and permits the statistical interpretation of groups of data elements.
- Feature completeness. Feature completeness refers to the degree to which all features of a given type for the area of the data set have been included.
- Lineage information. Information that describes processing tolerances, interpretation of rules applied to source materials, and basic production and quality assurance procedures. Lineage information should include all available information from the source.
- Logical consistency. Logical consistency refers to the fidelity of the relationships encoded in a data set. In a VPF data set, logical consistency requires that all topological foreign keys match the appropriate primitive, that all attribute foreign keys match the appropriate primitive or features, and that all tables described in the feature class schema tables maintain the relationships described.
- Positional accuracy. Positional accuracy refers to the root mean-square error (RMSE) of the coordinates relative to the position of the real-world entity being modelled. Positional accuracy must be specified without relation to scale and must contain all errors introduced by source elements, data capture, and data processing.

Web Sites on DIGEST

Following is a listing of web sites that provide or sell data in DIGEST format or maintain discussion forums on the format.

The Canadian Department of Defence

http://www.j2geo.ndhq.dnd.ca/digest/html/DIGEST.HTM

maintains an on-line copy of the DIGEST Standard. The Department of Defence, Defence Geomatics is also the custodian of the Standard. Other related DIGEST activities of interest can be found under the Canadian National Defence site. Among these is research and development efforts on DIGEST software tools at

http://www.j2geo.ndhq.dnd.ca/engr/projects/DGC/DGCBackground.htm

The Canadian site also contains a description of the Digital Geographic Information Working Group (DGIWG) which oversees the DIGEST efforts. The URL for the DGIWG description is located at

http://www.j2geo.ndhq.dnd.ca/defgeo/related/dgiwg/dgiwg.htm.

Information regarding the DIGEST Standard is also available at

http://www.nato.int/doc/standard.htm, Standardization Agreements (STANAG).

The U. S. National Imagery and Mapping Agency, http://www.nima.mil, maintains an on-line copy of the Vector Product Format Interface Standard (VPF). The document, along with other VPF-related documents and VPF product specifications, is located at

http://www.nima.mil/publications/specs/printed/vpf/vpf.html.

Activities related to the International Organization for Standardization Technical Committee 211, (ISO/TC 211), Geographic Information/Geomatics can be found at

http://www.statkart.no/isotc211 and http://www.opengis.org.

The Open GIS Consortium, Inc., an international consortium of private corporations, government agencies, non-government organizations and universities, coordinates the development and marketing of geoprocessing technologies.

A number of VPF or DIGEST datasets are available on-line too. The most widely available VPF dataset is the Digital Chart of the World (DCW). The Pennsylvania State University Digital Chart of the World Data Server

http://ortelius.maproom.psu.edu/dcw

gives worldwide coverage of DCW data. For data within the conterminous U.S., the DCW data are available from

http://h2o.er.usgs.gov/nsdi/dcw/dcwindes.html.

The University of Washington, through its China in Time and Space project, has an enhanced version of the Digital Chart of the World for China

http://citas.csdc.washington.edu/data/data.html

USGS also maintains a description of this dataset on-line

http://edcwww.cr.usgs.gov/glis/hyper/oldguides/dcw.

The Department of Mapping Sciences at the Agricultural University of Norway has an old research report on data quality issues regarding the Digital Chart of the World dataset

http://ilm425.nlh.no/gis/gis.html

This report still contains valuable information. This site has many useful links to other VPF-related sites also.

Commercial vendors that support VPF and DIGEST include Environmental Systems Research Institute (ESRI), Intergraph, Logiciels et Applications Scientifiques, and others.

ESRI provides VPF viewer software

http://www.esri.com/basc/products/arcview/arcview.html

and sells the DCW dataset in both VPF and non-VPF

http://www.esri.com/basc/data/catalog/esri/esri_dcw.htm.

Logiciels et Applications Scientifiques, Inc., http://www.las.com, works with the Canadian Department of Defence in the DIGEST Geo Components project to develop a set of geographic datastore interface for DIGEST's vector-based and non-vector-based products.

Intergraph has built a VPF data viewer and a VPF translator

http://www-nihon.intergraph.com/iss/products/mapping/translation/mge_vpfv.htm.

Nautical Data International, Inc.

http://www.ndi.nf.ca

is one of the commercial producers of geospatial data in VPF/DIGEST format. NDI is a distributor specializing in digital hydrographic products.

Index